Grain Legumes Towards Sustainable Agrifood Systems

NIPA® GENX ELECTRONIC RESOURCES & SOLUTIONS P. LTD.
New Delhi-110 034

About the Editors

Dr. Lalita Rana has been working as an Assistant Professor-*cum*-Scientist since 2019 in the Department of Agronomy Sugarcane Research Institute, Dr. Rajendra Prasad Central Agricultural University, Pusa, Bihar. She completed B.Sc. (Ag.) from Banaras Hindu University, Varanasi, Uttar Pradesh M.Sc. (Ag.) and Ph.D. in Agronomy from BCKVV, Mohanpur, West Bengal. With over 6 years of experience in teaching, extension and research, Dr. Rana has guided 5 students as major advisor and 8 students as co-advisor. She has published 30 research papers in national and international journals, 30 popular articles in farmer-friendly magazines, 10 book chapters, 2 authored books, 2 edited books, and 5 folders. She has received Young Women Scientist Award, Best Ph.D. Thesis award for her contributions to teaching and research field.

Dr. Navnit Kumar has been working as an Associate Professor-*cum*-Sr. Scientist since 2007 in the Department of Agronomy at the Sugarcane Research Institute, Dr. Rajendra Prasad Central Agricultural University, Pusa, Bihar. He completed B.Sc. (Ag.), M.Sc. (Ag.) and from RAU, Pusa, Bihar. With over 19 years of experience in extension, teaching and research, Dr. Kumar has guided 9 students as major advisor and 15 students as co-advisor. He has published 77 research papers in national and international journals, 81 popular articles in farmer-friendly magazines, 35 book chapters, 7 authored books, 8 edited books, and 13 folders.

Dr. Sumit Sow has completed Ph.D. in Agronomy with an ICAR-SRF from Dr. Rajendra Prasad Central Agricultural University, Pusa, Bihar. He has completed graduation from Visva-Bharati, West Bengal and M.Sc. in Agronomy with an ICAR-JRF Scholarship from Bihar Agricultural University, Sabour, Bihar. He has qualified ICAR-NET and UGC-NET. He has authored two textbooks and eight edited books. Additionally, he has published 22 research articles, 8 review articles, 35 popular articles, and 35 book chapters. He has received prestigious awards such as Best Poster, Best Article and Best M.Sc. Thesis Award.

Dr. Shivani Ranjan is currently working as Assistant Professor-cum-Scientist in the Department of Agronomy, Dr. Rajendra Prasad Central Agricultural University, Pusa, Bihar. She has completed B.Sc. (Ag.) and Ph.D. in Agronomy from Dr. Rajendra Prasad Central Agricultural University, Pusa, Bihar and M.Sc. in Agronomy from Bihar Agricultural University, Sabour, Bihar. She has qualified ICAR-NET, UGC-NET and was a recipient of UGC fellowship. She is one of the authors of the book "Extract of Agronomy" and the editor of nine books. She has published 20 research articles, 10 review articles, 34 popular articles and 30 book chapters. She has also received various awards like Best Poster, Best Article and Best Oral Presentation Award.

Mr. Ritwik Sahoo is currently pursuing Ph.D. in Plant Pathology from Uttar Banga Krishi Viswavidyalaya, Cooch Behar, West Bengal with DST INSPIRE Fellowship. He has completed graduation from UBKV, West Bengal and M.Sc. in Plant Pathology with ICAR-NTS Scholarship from Bihar Agricultural University, Sabour, Bihar and was University Gold Medalist in M.Sc. He has qualified ICAR ASRB-NET. He has published 2 research articles, 4 popular articles, 1 review article and 7 book chapters. He has received prestigious awards such as Best Poster and Best M.Sc. Thesis Award.

Dr. Adyant Kumar is working as Assistant Professor-*cum*-Junior Scientist in Department of Agronomy, Dr. Kalam Agricultural College, (Bihar Agricultural University), Kishanganj, Bihar. He had completed his Doctorate and Master degree in Agriculture with specialization in Agronomy from BCKV, West Bengal and RAU, Pusa (Bihar), respectively. He is the recipient of *Fakhruddin Ali Ahmed Award* for Outstanding Research in Tribal Farming Systems, 2021 given by ICAR under MoA&FW, GoI. He is also awarded with various awards & accolades *viz*. Best Oral Presentation, Best Thesis, Young Scientist and Young Achiever award as well as Valued Reviewer from Springer Nature etc. He performed as the Co-PI in the project "*Study of Millets production and consumption pattern in last 20 years in Jharkhand*" sanctioned by DRDM Tribal Welfare Research Institute, Ranchi. He has guided six M.Sc. students during his teaching period. He had written about 30 Research articles, 05 Review articles, 02 Book, 01 Souvenir, 10 Book chapters and 15 Popular articles in English and regional language for the academician, researcher, students and farming communities.

Grain Legumes Towards Sustainable Agrifood Systems

Lalita Rana
Navnit Kumar
Sumit Sow
Shivani Ranjan
Ritwik Sahoo
Adyant Kumar

NIPA® GENX ELECTRONIC RESOURCES & SOLUTIONS P. LTD.
New Delhi-110 034

NIPA® GENX ELECTRONIC RESOURCES & SOLUTIONS P. LTD.

101,103, Vikas Surya Plaza, CU Block
L.S.C. Market, Pitam Pura, New Delhi-110 034
Ph : +91-11-43860225, Mob.: +91 9717133558, 9540816132
E-mail: newindiapublishingagency@gmail.com
Website: www.nipaersources.com

Print ISBN: 978-93-58871-53-1
ebook ISBN: 978-93-58875-36-2

Composed and Designed by NIPA®.

Preface

The global agrifood landscape is undergoing rapid transformation, driven by an urgent need to enhance food security, promote environmental sustainability, and address the challenges posed by climate change. As the world population continues to grow, there is increasing pressure on agricultural systems to produce more food with fewer resources. Grain legumes-such as beans, lentils, peas, chickpeas, and soybeans-offer a vital solution to these challenges. They are not only rich sources of protein, vitamins, and minerals but also play a crucial role in sustainable farming practices through their ability to fix atmospheric nitrogen, improve soil health, and reduce greenhouse gas emissions.

The edited book entitled "**Grain Legumes towards Sustainable Agrifood Systems**" delves into the multifaceted roles that grain legumes play in shaping sustainable agrifood systems. This book brings together diverse perspectives from experts in the fields of agronomy, crop science, food technology, and policy-making to highlight the potential of grain legumes in achieving global food security and sustainability goals. Through a comprehensive exploration of topics such as genetic improvement, crop management, soil health, pest and disease management, value addition, and market development, it provides a holistic understanding of the current status, challenges, and future prospects of grain legumes in the context of sustainable agriculture.

The chapters in this book reflect the interdisciplinary nature of the subject, drawing on expertise from plant biology, agronomy, environmental science, nutrition, economics, and social sciences. Together, they provide a comprehensive overview of the potential of grain legumes to contribute to a sustainable future, highlighting both opportunities and challenges in realizing this potential. It covers a wide range of topics, from the genetic improvement of legumes for enhanced yield and stress resistance to innovative agronomic practices, value addition, and policy interventions that can promote the cultivation and consumption of these important crops.

We extend our gratitude to all contributors for their expertise and dedication to this volume. It is our hope that this book will not only enhance knowledge but also catalyze action towards a more sustainable, equitable, and nutritious food system for all.

We are thankful to NIPA, New Delhi for their interest and effort related to the publication of the book.

We believe that this book will serve as a valuable resource for researchers, agricultural practitioners, policymakers, students, and anyone interested in advancing sustainable food systems. By fostering a deeper understanding of the pivotal role of grain legumes, we hope to inspire further research, innovation, and collaboration to build a resilient and sustainable agrifood future.

Editors

Contents

1

Advance Molecular Approaches and Biotechnological Tools for Management of Pest and Diseases of Grain Legume Based Cropping System

Lalbiakzuali*

Department of Entomology, PGCA, Dr. Rajendra Prasad Central Agricultural University, Pusa, Bihar 848125, India

**Corresponding author's email: lalbiakzualiralteralte@gmail.com*

Abstract

Grain legumes face significant biotic stresses from various pests and pathogens, leading to substantial yield losses worldwide. This chapter synthesizes recent advances in molecular approaches and biotechnological tools for developing resistant varieties and managing diseases in legume-based cropping systems. Novel genomic technologies, including CRISPR-Cas9 gene editing, marker-assisted selection (MAS), and next-generation sequencing (NGS), have revolutionized our understanding of host-pathogen interactions and accelerated the development of resistant cultivars. RNA interference (RNAi) technology has emerged as a promising tool for targeted pest control, while transcriptomics and proteomics approaches have helped identify key resistance genes and defense-related proteins. Integration of these molecular tools with traditional breeding programs has resulted in improved varieties with enhanced resistance to major pests like pod borers and diseases such as fusarium wilt and ascochyta blight. Additionally, biotechnological applications including biopesticides and disease-forecasting systems have strengthened integrated pest management strategies. This review highlights the potential of combining multiple molecular approaches for sustainable pest and disease management in legume cultivation while identifying future research directions and challenges in their widespread adoption.

Keywords: *Biopesticides, Biotechnological tools, Marker-assisted selection, Legume*

Introduction

Grain legumes such as chickpea, pigeonpea, cowpea, field pea, lentil and *Phaseolus* beans are the principal source of dietary protein among vegetarians, and are an integral part of daily diet in several forms worldwide for human and animal consumption. However, the production and productivity of grain legumes is constrained by several biotic and abiotic factors, and suffer an average of 31.9 to 69.6 % loss in crop productivity due to insects and diseases (Chen *et al.*, 2011). Grain legumes, being a rich source of protein, are damaged by a large number of insect pests, of which gram pod borer (*Helicoverpa armigera*), spotted pod borer (*Maruca vitrata*), spiny pod borer (*Etiella zillckellella*), pod fly (*Melallagromyza obtuse*), stem fly (*Ophiomyia phaseoli*), aphids (*Aphis craccivora* and *Aphis fabae*), white fly (*Bemisia tabaci*) defoliators (*Spodoptera litura, S. exigua,* and *Amsacta spp.*) leafhoppers (*Empoasca spp.),* thrips (*Megaleurothrips dorsalis*, *Caliothrips illdicus*), blister beetles (*Mylabris spp.*) and the bruchids (*Collasobruchus chillellsis* and *Bruchus pisorum*) cause extensive losses (Sharma *et al.*, 2008). Major diseases are Fusarium wilts, rusts, powdery and downy mildews, ascochyta blights, botrytis gray molds, anthracnoses, damping-off, root rots, collar rots, vascular wilts and white molds (Rubiales *et al*., 2015). Pest management strategies in grain legumes require integration of different control tactics. Recent breakthroughs in molecular approaches and biotechnology offer innovative solutions. The application of modern tools of biotechnology, hybridization, genetic engineering, RNAi, CRISPR, molecular markers, quantitative trait loci (QTL) associated with insect and disease resistance can play a major role to accelerate the introgression of their resistance genes into high yielding cultivars (Sharma and Ortiz, 2002).

Molecular and Biotechnological Approaches

Marker-Assisted Breeding: The identification of molecular markers linked to the resistance genes, or if possible, located within the genes, could be used to replace or reduce screening tests in breeding populations, thus reducing the time and effort required to develop new cultivars. Molecular markers can be used to detect the individuals carrying resistance genes at seedling stages and avoid tedious inoculations and screenings. It can also be also used for pyramiding several genes conferring resistance to a disease, in order to increase the durability of the resistance conferred by each individual gene, or to combine genes (Rubiales *et al*., 2015). Many studies have been carried out to identify molecular markers linked to genes/QTLs controlling resistance to diseases and pest in legumes. Those include resistance to pests and soybean mosaic virus in soybean, bacterial blight, rust, virus, anthracnose, angular leaf spot, white mould and root rot in common bean. The efficiency of a marker in MAS depends on the distance between the flanking markers and the target gene that estimates the number of expected recombinants between the gene and the marker. In legumes the distance between the markers and the gene/QTL controlling resistance are in most cases still too large for efficient MAS.

In Chickpea, the preliminary linkage map based on interspecific crosses of *Cicer arietinum* × *C. reticulatum* and *Cicer arietinum* × *Cicer echinospermum* was made available by Gaur and Slinkard (1990). A mapping population of 126 F13 RILs of ICCV 2 × 1G62, has been evaluated for resistance to *Helicoverpa armigera*. Mapping population from the cross between Vijay (susceptible) × ICC 506EB (resistant) has also been evaluated for resistance to *H. armigera*. Panguluri *et al.* (2006) used AFLP markers to detect polymorphism in cultivated pigeonpea and two of its wild relatives *Cajanus volubilis* and *Rhynchosia bracteata*. High levels of resistance to pod borer, *H. armigera*, and pod fly, *M. obtusa*, have been identified in wild relatives of pigeonpea such as *C. scarabaeoides*, *C. sericeus*, and *C. acutifolius*. The first genetic linkage map of cultivated groundnut contained 350 RFLP loci distributed across 22 linkage groups, with a total map distance of approximately 2,700 cM (Burow et at. 1999). RAPD (RKN 229, RKN 410, and RKN 440) and RFLP (R2430E, R2545E, and Sl137E) markers linked with resistance to root-knot nematode has also been identified.

Quantitative Trait Locus (QTLs): While complete or high levels of resistance conferred by single genes can be easily introduced into varieties by backcrossing, the introgression of quantitative resistance governed by several minor genes/QTLs is challenging. A Quantitative Trait Locus (QTL) is defined as a genetic region that influences the variation of a quantitative trait, with its detection influenced by factors such as trait heritability, marker density, mapping algorithm, and sample size. In monogenic resistances, the individuals carrying the gene are easily identified in breeding populations while quantitative polygenic traits are usually highly influenced by the environment and the genotype cannot be directly inferred from the phenotype. As a result, most resistant varieties available so far are based on major genes although some successes have been obtained in some polygenic resistances (Rubiales *et al.*, 2015). QTLs can be used to breed plants that are more resistant to disease and pests, and to develop more effective and durable resistance. The development of QTL analysis was a major breakthrough in the characterisation of the inheritance of quantitative traits, enabling the identification of important genomic regions and their contribution to phenotypic variation. In addition, the mapping of QTLs is a useful tool to identify molecular markers linked to the resistance genes that could be used to assist breeding or to identify the gene itself (Torres *et al.*, 2006).

Some examples of QTLs in disease and pest control include:

- A study identified QTLs associated with resistance to pea weevil (*Bruchus pisorum*) in a genetic linkage map of pea. The study found that QTLs were associated with reduced seed infestation and larval development, providing an opportunity for breeders to generate effective and sustainable strategies for weevil control (Aznar *et al.*, 2020).
- A QTL explaining 63 % of the resistance to *Uromyces pisi* has been identified flanked by 2 RAPDs (Random Amplification of Polymorphic DNA) (Barilli *et al.*, 2012).
- Only incomplete resistance was identified in chickpea against *Uromyces ciceris-arietini*. A QTL explaining 81% of the resistance has been described and assigned hypothetically to a single gene (Madrid *et al.*, 2008).
- In peanut, genetic studies indicated that resistance to the rust fungus *Puccinia arachidis* is complex and polygenic, with 12 QTLs reported (Khedikar *et al.*, 2010).
- Complete or high level of resistance to *Erysiphe pisi*, conferred by the genes ER1, ER2 or ER3 is available in pea (Fondevilla *et al.*, 2008).
- Resistance to *Ralstonia solanacearum* in *Medicago truncatula* is governed by major QTL located on chromosome 5 (Vailleau *et al.*, 2007).
- Resistances to pea enation mosaic virus, bean yellow mosaic and bean leaf roll virus have also been identified in faba bean (Kumari and Makkouk, 2003).

In Vitro–Based Breeding Approaches

Tissue culture and its derived *in vitro* approaches are an important complement to traditional breeding methods. However, all these methods are dependent on efficient *in vitro* regeneration to recover fertile plants from tissue explants or isolated protoplasts, which have proven to be arduous in legumes. Indeed, most legumes are recalcitrant to regeneration from both organogenesis and embryogenesis. This recalcitrance is the main bottleneck against the routine production of transgenic plants for many legumes, for both new cultivar releases and functional genomic studies since advances in molecular genetics, i.e. manipulation of endogenous gene expression or insertional mutagenesis, require efficient transformation systems (Choudhary *et al.*, 2009). Thus many efforts have been made to optimize protocols of *in vitro* regeneration in many legumes including alfalfa, common bean, cowpea, chickpea, grass pea, lentil. Although in many cases, the recovery of fertile plants is still low, it allowed application of genetic transformation and other tissue culture derived techniques to generate genetic diversity such as somaclonal variation, in vitro mutagenesis, and wide hybridization in some legume species. The advances gained in vitro related approaches to improve legume genetic variability and their application to improve crop resistance to diseases and pests (Rubiales *et al.*, 2015).

Wide Hybridization: Wide hybridization or interspecific hybridization corresponds to the creation of a hybrid between plants from different species and/or genera. As such it offers the breeder the possibility to broaden the genetic variability outside the species range and to take advantage of the large genetic variability found in related wild species that often contain interesting traits such as resistance to diseases and pests. While hybridization of two closely related species may provide fertile plants, in most cases these species are completely or partially sexually incompatible leading to sterile hybrids. Thus application of wide hybridization methods offers the theoretical possibility to transfer and introgress resistance traits from wild species to legume crops. Introgression of resistance to a number of pests and diseases has been described in chickpea, common bean, lentil, pea, soybean, white clover, *Vigna spp.* and pigeonpea. Anthracnose resistance has successfully been transferred from the wild species *Lens ervoides* to lentil through embryo rescue (Fiala *et al.,* 2009). In addition, the *L. culinaris* × *L. ervoides* lines may possess higher degree of resistance to *Stemphylium botryosum* (Tullu *et al.,* 2011). Resistance to ascochyta blight in chickpea may be introgressed from *C. reticulatum* and *C. echinospermum* (Collard *et al.,* 2003). Resistance to several races of the cyst nematode *Heterodera glycine* was also conferred to soybean by crossing a *Glycine max* cultivar to one accession of its closely related *Glycine tomentella.* Other resistances were more difficult to transfer such as the resistance to *Phytophthora drechsleri* f.sp. *cajani* that was recovered from crosses of pigeonpea with the tertiary gene pool species *C. platycarpus* that were made using embryo rescue. The progenies also showed potential for conferring resistance to several insect pests including pod borer and bruchid (Mallikarjuna *et al.,* 2011). Transfer of aphid and weevil resistance to fertile S1 generation of alfalfa was obtained by protoplast fusion of *Medicago rugosa* × *M. sativa*. Resistance to various aphid species has been transferred from wild accessions to *M. truncatula*, which in addition to being a model legume is an annual pasture crop in parts of the world (Mizukami *et al.*, 2006).

Genetic Engineering: Many if not all legume species can be genetically transformed including pea, chickpea, soybean, *M. truncatula*, peanuts, common bean and cowpea with limited recovery rates in most cases. Legume transformation has been mainly performed by *Agrobacterium tumefaciens* but in some cases, especially in common bean, micro-particle bombardment has been used to deliver exogenous DNA to embryogenic or organogenic tissues (Chandra and Pental, 2003). Despite these efforts, in most cases the recovery rate of transgenic lines is still low and about 1% depending on the transformation method and the legume species. The recovery rate is generally ascribed to the recalcitrant nature of most legumes to in vitro regeneration, which is the main bottleneck to the application of genetic transformation in legumes. By contrast to induced-mutagenesis that targeted mainly fungal diseases, genetic transformation was mainly aimed to confer resistance to insect and virus diseases.

Thus one of the main transformation strategies was the introduction of derivatives of *cry1* genes from *Bacillus thuringiensis* conferring resistance to many pod borer insects. To improve the efficiency of resistance, the strategy has moved from the transfer of a single *cry1* gene to constructs containing cry genes with different modes of action. So far, the transgenic lines of pigeonpea, chickpea, cowpea and soybean resistant to the lepidopterans *Helicoverpa armigera, Spodoptera litura, Maruca vitrata, Helicoverpa zea, Anticarsia gemmatalis* and *Pseudoplusia includens*, respectively, contain only single cry genes, although in some cases an option exists for stacking two different genes by crossing. Expression of cry3a in alfalfa provides resistance to *Hypera postica*, the alfalfa weevil (Tohidfar *et al.*, 2013). Alternatively, high levels of resistance to different bruchids was also shown in transgenic pea, chickpea, adzuki bean and cowpea expressing high levels of common bean α-amylase 1 inhibitor in seeds. Although several strategies have been used to confer resistance to viruses in legumes, the most frequently used method consists of expressing viral coat or other viral genes within the host. This resulted in protection of pea against pea enation mosaic virus and alfalfa mosaic virus.

Table 1. Some genetically engineered legumes for biotic stress

Legume target	Biotic stress	Gene(s)
Cajanus cajan	*Spodoptera litura* *Helicoverpa armigera*	Cry1 E-C Cry1 AcF
Glycine max	*Sclerotinia sclerotiorum* *Helicoverpa zea*	Oxalate oxidase (Germin (gf-2.8) fom wheat) *cry1Ac*
Medicago sativa	*Phoma medicaginis*	Resveratrol synthase from peanut
Pisum sativum	*Bruchus pisorum*	Alpha-amylase inhibitor
Vigna unguiculata	*Maruca vitrata*	*Cry1Ab*
Vigna angularis	*Callosobruchus chinensis*	Alpha-amylase inhibitor from bean

RNAi: RNAi is a powerful tool for protecting agricultural crops from pest damage. RNAi can be initiated by feeding double-stranded RNA (dsRNA) to the target insect pest. These dsRNAs are cut within cells into short interfering RNAs (siRNAs) by Dicer enzymes. The siRNAs, carried by a protein complex, scan for other RNAs with a complementary sequence. These target sequences are bound by the siRNA and subsequently cleaved, thereby preventing the production of proteins essential for survival of the insect pest. RNAi has been successfully applied for management of insects and diseases (Nidheesh *et al.,* 2022). Although RNAi works well against beetles (order coleoptera), it does not work well against other groups of insects, including moths (order lepidoptera).

RNA-interference (RNAi) strategy has emerged as an efficient means to control begomoviruses infection in crops including legumes RNAi-derived transgenic resistance has been accomplished by targeting the AC1 to control begomoviruses like Bean golden mosaic virus (Kumar *et al.*, 2017). Genetically modified plants expressing hairpin RNA (hpRNA) were constructed to suppress the CYP6AE14 expression in cotton bollworm. For example, RNAi can be achieved by feeding insects with dsRNA-producing *E. coli.* Feeding insects with dsRNA-producing bacteria could lead to a more sustained release of the dsRNA in the insect and could help avoid rapid degradation in the digestive system (Christiaens *et al.,* 2020).

CRISPR: CRISPR/Cas9 (Clustered Regularly Interspaced Short Palindromic Repeats-Cas9) is a revolutionary gene-editing tool derived from the bacterial defense system against viruses. It enables precise modifications to an organism's DNA (Kumar *et al.*, 2023). CRISPR/Cas9 can be used to modify insects and pathogens to prevent them from feeding on plants, or to modify plants to repel insects. CRISPR-Cas gene editing has been effectively used over the past ten years to create insect-resistant plants and to modify a number of insects. It renders the pests infertile, reducing populations, disables genes essential for pest survival, reproduction or virulence or introduces genes producing toxins harmful to pests and diseases.

Chickpea is considered recalcitrant to in vitro tissue culture amongst all edible legumes. The clustered, regularly interspaced short palindromic repeats/ CRISPR-associated protein 9 (CRISPR/Cas9)-based genome editing in chickpea can remove the bottleneck of limited genetic variation The CRISPR-Cas method's deletion of the *Orco* gene from *Spodoptera litura* affected its choice of a mating partner and host plant (Koutroumpa *et al.,* 2016). CRISPR/ CasRx antiviral systems have been used to develop soybeans with increased resistance to the soybean mosaic virus (SMV) (Gao *et al.* 2024).

Conclusion

Diseases and pests are the main causes of yield reduction of legume impairing both yield and quality of the crops. Although development of genomic resources in legumes has been relatively slow compared to other crops such as cereals, the wide applicability of molecular and biotechnological tools has already been demonstrated in various pulse crops but are still in early infancy stage for most legumes. This will no doubt contribute to unraveling key components of legume defense and resistance to pathogens and pests in the near future. At this stage, biotechnological approaches such as marker-assisted selection, tissue culture, in vitro mutagenesis, genetic transformation, RNAi and CRISPR technology could contribute to speeding up classical breeding and overcome major problems such as lack of natural sources of resistance and sexual incompatibility for more efficient crop improvement. The progress in deploying these resources, especially to legumes, has been depressingly slow. However, it is atask of breeders, entomologists and pathologists to produce cultivars with insect and disease resistance to increase crop productivity and improve livelihoods of the rural poor.

References

Aznar-Fernández, T., Barilli, E., & Cobos, M. J. (2020). Identification of quantitative trait loci (QTL) controlling resistance to pea weevil (Bruchus pisorum) in a high-density integrated DArTseq SNP-based genetic map of pea. Scientific Reports, 10, 33.

Barilli, E., Rubiales, D., & Castillejo, M. A. (2012). Comparative proteomic analysis of BTH and BABA-induced resistance in pea (Pisum sativum) toward infection with pea rust (Uromyces pisi). Journal of Proteomics, 75, 5189-5205.

Burow, M. D., Simpson, C. E., Starr, J. L., & Paterson, A. H. (1999). Generation of a molecular marker map of the cultivated peanut, Arachis hypogaea L. In Proceedings of the Plant and Animal Genome VII, pp. 17-21.

Chandra, A., & Pental, D. (2003). Regeneration and genetic transformation of grain legumes: An overview. Current Science, 84, 381-387.

Chen, W., Sharma, H. C., & Muehlbauer, F. (Eds.). (2011). Compendium of chickpea and lentil diseases and pests. The American Phytopathological Society, p 39.

Choudhary, K., Singh, M., Rathore, M. S., & Shekhawat, N. S. (2009). Somatic embryogenesis and in vitro plant regeneration in moth bean [Vigna aconitifolia (Jacq.) Marechal]: A recalcitrant grain legume. Plant Biotechnology Reports, 3, 205-211.

Christiaens, O., Whyard, S., Vélez, A. M., & Smagghe, G. (2020). Double-stranded RNA technology to control insect pests: Current status and challenges. Frontiers in Plant Science, 11, Article 451.

Collard, B. C. Y., Pang, E. C. K., Ades, P. K., & Taylor, P. W. J. (2003). Preliminary investigation of QTLs associated with seedling resistance to ascochyta blight from Cicer echinospermum, a wild relative of chickpea. Theoretical and Applied Genetics, 107, 719-729.

Fiala, J. V., Tullu, A., Banniza, S., Seguin-Swartz, G., & Vandenberg, A. (2009). Interspecies transfer of resistance to anthracnose in lentil (Lens culinaris Medic.). Crop Science, 49, 825-830.

Fondevilla, S., Rubiales, D., Moreno, M. T., & Torres, A. M. (2008). Identification and validation of RAPD and SCAR markers linked to the gene Er3 conferring resistance to Erysiphe pisi DC in pea. Molecular Breeding, 22, 193-200.

Gao, L., Xie, L., Xiao, Y., Cheng, X., Pu, R., Zhang, Z., Liu, Y., Gao, S., Zhang, Z., Qu, H., & Zhi, H. (2024). CRISPR/CasRx-mediated resistance to Soybean mosaic virus in soybean. The Crop Journal, 12(4), 1093-1101.

Gaur, P. M., & Slinkard, A. E. (1990). Genetic control and linkage relations of additional isozyme markers in chickpea. Theoretical and Applied Genetics, 80, 648-656.

Khedikar, Y. P., Gowda, M. V. C., Sarvamangala, C., Patgar, K. V., Upadhyaya, H. D., & Varshney, R. K. (2010). A QTL study on late leaf spot and rust revealed one major QTL for molecular breeding for rust resistance in groundnut (Arachis hypogaea L.). Theoretical and Applied Genetics, 121, 971-984.

Koutroumpa, F. A., Monsempes, C., François, M. C., de Cian, A., Royer, C., Concordet, J. P., & Jacquin-Joly, E. (2016). Heritable genome editing with CRISPR/Cas9 induces anosmia in a crop pest moth. Scientific Reports, 6, 29620.

Kumar, S., Tanti, B., Patil, B. L., Mukherjee, S. K., & Sahoo, L. (2017). RNAi-derived transgenic resistance to Mungbean yellow mosaic India virus in cowpea. PLOS ONE, 12(10), Article e0186786.

Kumari, S. G., & Makkouk, K. M. (2003). Differentiation among Bean leaf roll virus susceptible and resistant lentil and faba bean genotypes on the basis of virus movement and multiplication. Journal of Phytopathology, 151, 19-25.

Kumar, A., Singh, R., Singh, D., Satyendra, Sow, S., Ranjan, S., & Singh, S. K. (2023). Gene Pyramiding: A Molecular Approaches to Overcome Biotic and Abiotic Stresses. In: Innovative Approaches in Plant Protection, Prakhar Goonj Publication, pp. 319-332.

Madrid, E., Rubiales, D., Moral, A., Moreno, M. T., Millán, T., Gil, J., & Rubio, J. (2008). Mechanism and molecular markers associated with rust resistance in a chickpea interspecific cross (Cicer arietinum L. × Cicer reticulatum Lad.). European Journal of Plant Pathology, 121, 43-53.

Mallikarjuna, N., Senapathy, S., Jadhav, D. R., Saxena, K., Sharma, H. C., Upadhyaya, H. D., Rathore, A., & Varshney, R. (2011). Progress in the utilization of Cajanus platycarpus (Benth.) Maesen in pigeonpea improvement. Plant Breeding, 130, 507-514.

Mizukami, Y., Kato, M., Takamizo, T., Kanbe, M., Inami, S., & Hattori, K. (2006). Interspecific hybrids between Medicago sativa L. and annual Medicago containing alfafa weevil resistance. Plant Cell, Tissue and Organ Culture, 84, 79-88.

Nidheesh, T. D., Jagadish, K. S., Preethi, C., Arunkumar, C. G., & Prabhulinga, T. (2022). Recent genetic tools for the management of stored product pests. In Genetic methods and tools for managing crop pests Springer, pp. 559-567.

Panguluri, S., Janaiah, K., Govil, J., Kumar, P., & Sharma, P. (2006). AFLP fingerprinting in pigeon pea (Cajanus cajan (L.) Millsp.) and its wild relatives. Genetic Resources and Crop Evolution, 53, 523-531.

Rubiales, D., Fondevilla, S., Chen, W., Gentzbittel, L., Higgins, T. J., Castillejo, M. A., Singh, K. B., & Rispail, N. (2015). Achievements and challenges in legume breeding for pest and disease resistance. Critical Reviews in Plant Sciences, 34(1-3), 195-236.

Sharma, H. C., & Ortiz, R. (2002). Host plant resistance to insects: An eco-friendly approach for pest management and environment conservation. Journal of Environmental Biology, 23, 11-35.

Sharma, H. C., Varshney, R. K., Gaur, P. M., & Gowda, C. L. L. (2008). Potential for using morphological, biochemical, and molecular markers for resistance to insect pests in grain legumes. Journal of Food Legumes, 21(4), 211-217.

Tohidfar, M., Zare, N., Jouzani, G. S., & Eftekhari, S. M. (2013). Agrobacterium mediated transformation of alfalfa (Medicago sativa) using a synthetic cry3a gene to enhance resistance against alfalfa weevil. Plant Cell, Tissue and Organ Culture, 113, 227-235.

Torres, A. M., Román, B., Avila, C., Satovic, Z., Rubiales, D., Sillero, J. C., Cubero, J. I., & Moreno, M. T. (2006). Faba bean breeding for resistance against biotic stresses: Towards application of marker technology. Euphytica, 147, 67-80.

Tullu, A., Diederichsen, A., Suvorova, G., & Vandenberg, A. (2011). Genetic and genomic resources of lentil: Status, use and prospects. Plant Genetic Resources: Characterization and Utilization, 9, 19-29.

Vailleau, F., Sartorel, E., Jardinaud, M. F., Chardon, F., Genin, S., Huguet, T., Gentzbittel, L., & Petitprez, M. (2007). Characterization of the interaction between the bacterial wilt pathogen Ralstonia solanacearum and the model legume plant Medicago truncatula. Molecular Plant-Microbe Interactions, 20, 159-167.

2

Biological Nitrogen Fixation Mechanism in Grain Legumes An Overview

Manish Kumar[1], Shweta Kumari[1*], Vandana Kumari[1], Rina Roy[2] and Sweeti Kumari[2]

[1]*Department of Soil Science and Agricultural Chemistry, Dr. Kalam Agricultural College, Kishanganj, Bihar Agricultural University, Sabour, Bihar, India*

[2]*Department of Agronomy, Dr. Kalam Agricultural College, Kishanganj, Bihar Agricultural University, Sabour, Bihar, India*

**Corresponding author's email: shewtacsa11oct@gmail.com*

Abstract

Biological nitrogen fixation (BNF) is a critical ecological process carried out by grain legumes, which form symbiotic associations with nitrogen-fixing bacteria, primarily Rhizobium species. This natural mechanism converts atmospheric nitrogen (N_2) into ammonia, a form that plants can readily utilize, significantly enhancing soil fertility and reducing the need for synthetic nitrogen fertilizers. The process begins with the recognition and infection of the legume roots by rhizobia, leading to the formation of specialized structures called nodules, where the bacteria convert atmospheric nitrogen into ammonia through the enzymatic action of nitrogenase. The symbiosis between legumes and rhizobia is tightly regulated by various genetic and environmental factors, including plant signalling molecules, soil conditions, and nutrient availability. Grain legumes, such as soybeans, chickpeas, and lentils, play a vital role in sustainable agriculture by improving soil health, promoting biodiversity, and providing a protein-rich food source. This overview explores the molecular mechanisms underlying BNF, the factors influencing the efficiency of this process, and its potential to enhance sustainable agricultural practices. Understanding the intricacies of BNF in grain legumes can inform breeding programs and management strategies aimed at maximizing the benefits of this natural process for agricultural productivity and environmental sustainability.

Keywords: *Nitrogenase, Biological Nitrogen Fixation, Environmental sustainability, Soil fertility*

Introduction

Nitrogen is an essential nutrient required for the growth and development of plants, and it is a crucial component of many biomolecules, such as proteins, nucleic acids, and chlorophyll. While nitrogen is abundant in the atmosphere, accounting for 78% of the air we breathe, most plants are unable to directly utilize this atmospheric nitrogen. Instead, they rely on the process of biological nitrogen fixation, which is carried out by a specialized group of microorganisms called diazotrophs. Biological nitrogen fixation (BNF) is a process by which certain microorganisms, including bacteria and archaea, convert atmospheric nitrogen into ammonia (NH_2), a form that plants can readily assimilate. Grain legumes, such as soybeans, peanuts, and lentils, are among the most important crops that can engage in this symbiotic relationship with nitrogen-fixing bacteria. This symbiosis not only fulfils the nitrogen needs of the legumes themselves but also contributes to the nitrogen economy of agricultural systems, reducing the need for synthetic nitrogen fertilizers and promoting sustainable agriculture. The process of biological nitrogen fixation in grain legumes is mediated by Rhizobia, a group of soil bacteria that form specialized structures called root nodules on the plants roots. Within these nodules, the *Rhizobia* bacteria convert atmospheric nitrogen (N_2) into a form that can be readily absorbed and utilized by the plant, known as ammonia. The nitrogen fixation process in legumes involves a complex series of steps, starting with the initial infection of the plant's roots by *Rhizobia* bacteria (Sow *et al.*, 2023). The bacteria release chemical signals that trigger the formation of root nodules, where they can establish a symbiotic relationship with the plant.

The total amount of nitrogen fixed usually ranges from 100 to 300 kg N/ha depending on factors such as legume species (and cultivar), length of growing season and environmental conditions. The grain legumes account for 27% of the world crop production and are only next to cereals. Globally, 60% of the total human protein requirement is met through plant-based proteins, of which legumes contribute to about 50% (Smykal *et al.*, 2015; Leinonen *et al.*, 2019). An assessment of global protein sustainability favors a shift from animal-based to plant-based proteins (Chaudhary *et al.*, 2018; Leinonen *et al.*, 2019). The legumes are valuable in agriculture for two main reasons. One, the legume seeds are rich in protein content (20–25% of the dry weight) and help meet the daily protein requirement of a majority of the world population, especially for those who cannot afford to source it from animals. Second, the environmental cost of legume cultivation, calculated as life cycle assessment (LCA), is significantly lower compared to non-legume crops due to the former's ability to fix their own nitrogen (MacWilliam *et al.*, 2014).

Significance of Biological Nitrogen Fixation in Agriculture

Nitrogen is often the most limiting nutrient in agricultural systems, and its availability directly affects crop yield and quality. Traditional agriculture has relied heavily on synthetic nitrogen fertilizers, which are energy-intensive to produce and can lead to environmental problems such as water pollution, greenhouse gas emissions, and soil degradation. In contrast, biological nitrogen fixation offers a natural and sustainable alternative. Grain legumes play a crucial role in this process through their symbiotic relationship with rhizobia, which fix atmospheric nitrogen into ammonia within specialized root structures called nodules. This fixed nitrogen not only supports the growth and development of the legume itself but also enriches the soil for subsequent crops in a rotation system. As such, legumes are often used in crop rotations to enhance soil fertility and reduce the need for chemical fertilizers. Furthermore, the nitrogen fixed by legumes can contribute to improved soil structure, increased microbial activity, and better water retention, all of which are beneficial for long-term soil health. This makes BNF an integral component of sustainable agricultural practices, particularly in low-input and organic farming systems (Ranjan *et al.*, 2023).

The annual demand of chemically synthesized nitrogen is projected at about 120 T (FAO, 2017), approximately 50% of which is required to fertilize wheat, maize, and rice (Ladha *et al.*, 2016). The N-use efficiency for cereals is estimated to be about 33%; the remaining becomes either a source of aquatic systems pollution (Raun and Johnson, 1999) or a substrate for N_2O or nitrous oxide (NO) gases in the environment through denitrification. According to the US Environmental Protection Agency (EPA), N_2O has 294 times more GHG effect than carbon dioxide on a per unit mass basis. In addition, the process of nitrogen fertilizer synthesis uses fossil fuels as one of the energy sources. Depending on the type of the fossil fuel, it releases 1.10–3.37 t of CO_2/ton fossil fuel burnt. It has been predicted that 2% consumption of global energy by 2050 will be due to chemical synthesis of N fertilizers (Glendining *et al.*, 2009). Approximately, one-third of total GHG emission occurs via the agriculture sector including crop cultivation and livestock (Gilbert, 2012). A large carbon footprint of agriculture along with the detrimental effects of runoff fertilizers puts a question mark on agriculture sustainability. The need for minimizing GHG emission and protecting the climate has been highlighted in a report by the Intergovernmental Panel on Climate Change of 2011. In this connection, the legume production systems hold a great promise to reduce the agriculture carbon footprint by expanding the acreage under legumes and making the SNF process more efficient.

Mechanism of Nitrogen Fixation in Grain Legumes

Biological nitrogen fixation (BNF), discovered by Beijerinck in 1901, is carried out by a specialized group of prokaryotes. These organisms utilize the enzyme nitrogenase to catalyze the conversion of atmospheric nitrogen (N_2) to ammonia (NH_3). Plants can readily assimilate NH_3 to produce the aforementioned nitrogenous biomolecules. These prokaryotes include aquatic organisms, such as cyanobacteria, free-living soil bacteria, such as *Azotobacter*, bacteria that form associative relationships with plants, such as *Azospirillum*, and most importantly, bacteria, such as *Rhizobium* and *Bradyrhizobium*, that form symbioses with legumes and other plants (Postgate, 1982). The process of nitrogen fixation in grain legumes is complex and involves a well-orchestrated sequence of events, from the initial recognition between the legume and rhizobia to the final assimilation of fixed nitrogen into organic molecules.

Symbiotic Association Between Legumes and Rhizobia: Many microorganisms fix nitrogen symbiotically by partnering with a host plant. The plant provides sugars from photosynthesis that are utilized by the nitrogen-fixing microorganism for the energy it needs for nitrogen fixation. In exchange for these carbon sources, the microbe provides fixed nitrogen to the host plant for its growth. Even though the symbiotic partners described above play an important role in the worldwide ecology of nitrogen fixation, by far the most important nitrogen-fixing symbiotic associations are the relationships between legumes and *Rhizobium* and *Bradyrhizobium* bacteria. Important legumes used in agricultural systems include alfalfa, beans, clover, cowpeas, lupines, peanut, soybean, and vetches. Of the legumes in agricultural production, soybeans are grown on 50% of the global area devoted to legumes, and represent 68% of the total global legume production (Vance, 2001).

The symbiotic relationship between grain legumes and rhizobia is initiated by a sophisticated signalling process between the plant roots and the bacteria. The interaction begins when legumes secrete specific signalling molecules called flavonoids into the rhizosphere. These flavonoids are detected by rhizobia, triggering the production of lipo-chitooligosaccharides known as Nod factors.

Nodulation Process

- **Root Hair Curling:** The Nod factors bind to receptors on the surface of legume root hairs, causing them to curl and entrap the rhizobia what is called a shepherd's crook.
- **Infection Thread Formation:** The bacteria penetrate the root hair through a localized degradation of the cell wall, forming an infection thread-a tubular structure that guides the bacteria through the root cells.

- **Nodule Formation:** As the infection thread reaches the root cortex, the rhizobia are released into the cortical cells, where they induce cell division, leading to the formation of nodules. These nodules are specialized organs where nitrogen fixation takes place.

Within the nodules, rhizobia differentiate into a nitrogen-fixing form called bacteroids. The plant supplies the bacteroids with carbohydrates derived from photosynthesis, which the bacteroids use to generate ATP and reducing power (in the form of NADH) required for the nitrogen fixation process.

The Interaction Between the Bacteria and Host Legume: More than 98 species of symbiotic nitrogen-fixing rhizobia exist in 14 taxa in association with legumes (Berrada and Fikri-Benbrahim, 2014). Many legumes have evolved to establish a symbiosis with nitrogen-fixing soil-bacteria collectively known as rhizobia (including the genera *Azorhizobium, Allorhizobium, Bradyrhizobium, Mesorhizobium, Rhizobium* and *Sinorhizobium*). The interaction between the bacteria and host legume is so intricate that a particular *Rhizobium* or *Bradyrhizobium* will only nodulate a select number of plant genera. For example, *Rhizobium melilotii* will only nodulate alfalfa, while *Rhizobium leguminosarum biovar trifolii* will only nodulate clover (*Trifolium*). This host specificity is referred to cross inoculation group cell signalling between the bacteria and the legume host. The aforementioned Nod factors have been identified as lipo-chitionoligosaccharides. Variations in the structures of these oligosaccharides determine the host specificity for the bacterium. Major cross inoculation groups are listed in Table 1.

Table 2.1: Major cross inoculation groups of legumes

Crop	Nodulating Bacteria
Alfalfa	*Sinorhizobium melilotii*
Beans	*Rhizobium legumninosarum biovar phaseoli and Rhizobium tropici*
Clover	*Rhizobium leguminosarum biovar trifolii*
Lotus	*Mesorhizobium loti*
Peas	*Rhizobium leguminosarum biovar viceae*
Soybean	*Bradyrhizobium japonicum, Bradyrhizobium elkanii, Rhizobium fredii*
Sesbania	*Azorhizobium caulinodans*

Biochemical Pathway of Nitrogen Fixation: The actual process of nitrogen fixation occurs within the bacteroids housed in the nodule cells. This process is catalyzed by the nitrogenase enzyme complex, which consists of two main components: the Fe protein (Dinitrogenase reductase) and the Mo-Fe protein (Dinitrogenase).

Key Components of the Nitrogenase Complex

- **Fe Protein (Dinitrogenase Reductase):** This component has two identical subunits, each containing an iron-sulfur cluster. The Fe protein is responsible for transferring electrons from a low-potential electron donor, such as ferredoxin or flavodoxin, to the MoFe protein. This electron transfer is coupled with ATP hydrolysis.
- **MoFe Protein (Dinitrogenase):** This is a larger, heterotetrameric protein containing two subunits, each with a molybdenum-iron cofactor (Mo-Fe -cofactor) that is the active site for nitrogen reduction. The Mo-Fe protein catalyzes the conversion of atmospheric nitrogen (N_2) into ammonia (NH_3).

Nitrogen Fixation Reaction: The overall reaction catalyzed by nitrogenase can be summarized as follows:

$N_2 + 8H^{+} + 8^{e-} + ATP \rightarrow 2NH_3 + H_2 + 16AD + 16P_i$

In the mechanism of biological nitrogen fixation in legumes, two nitrogen molecules are combined with two hydrogen molecules in the presence of 8 electrons, 8 protons, and 16 ATPs. Then, we get 2 ammonia molecules, 16 ADP (adenosine diphosphate) molecules, 16 inorganic phosphate, and two hydrogen molecules (Hubbell and Kidder, 2009). This reaction requires a significant amount of energy, provided by the hydrolysis of 16 ATP molecules per molecule of nitrogen reduced. The process also produces hydrogen gas (H_2) as a byproduct, which can be recycled by hydrogenase enzymes within the bacteroids to produce additional ATP.

Oxygen Regulation in Nodules: The nitrogenase enzyme is highly sensitive to oxygen, which can irreversibly inactivate the enzyme by oxidizing its iron-sulfur clusters. However, the bacteroids require oxygen for respiration to generate the ATP needed for nitrogen fixation. This creates a paradox that is resolved by the plant through the production of leghemoglobin, an oxygen-binding protein that is abundant in the nodules. Leghemoglobin controls the concentration of free oxygen in the cytoplasm of infected plant cells, protecting nitrogenase from oxygen while at the same time enabling the provision of oxygen for respiration in root tissue to supply the energy required. A fascinating part of this is leghemoglobin is closely related to the hemoglobin in blood with an analogous function in transporting oxygen. Like hemoglobin, leghemoglobin is red when charged with oxygen. This explains why healthy root nodules are pink. The presence of a large number of nodules that are pink when split open is a reliable indicator of successful establishment of BNF in legumes crops (Fig.2.1).

Fig. 2.1: A photo of a split nodule showing the characteristic pink colour. (www.gartensoja.de)

Role of Leghemoglobin

- **Oxygen Transport:** Leghemoglobin binds oxygen with high affinity, controlling the oxygen concentration within the nodule to ensure that it is low enough to protect the nitrogenase enzyme but sufficient to support bacteroid respiration.
- **Oxygen Buffering:** By binding and releasing oxygen as needed, leghemoglobin acts as an oxygen buffer, maintaining a stable microenvironment within the nodule.

The presence of leghemoglobin gives the nodules their characteristic pink or red color, which is often used as an indicator of active nitrogen fixation. The nodules are rich in iron and protein providing a rich source of food for larvae of certain weevils (*Sitona lineatus* and other *Sitona* species). The leghemoglobin is also so similar to mammalian blood that it is used in substitute meat products. Legumes plants form two types of nodules: indeterminate ovoid shaped and determinate round shaped (Fig. 2.2).

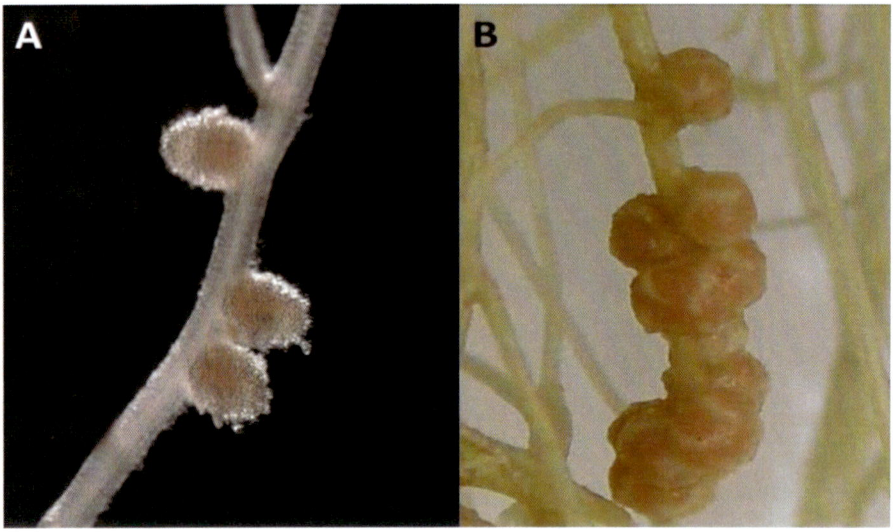

Fig. 2.2: Shape of nodules in legume plants (A- indeterminate; B-determinate)

Ammonia Assimilation: Once ammonia (NH_2) is produced by nitrogenase, it must be rapidly assimilated into organic forms to prevent its accumulation, which can be toxic to the plant. Ammonia is first converted into glutamine and then into other amino acids through the glutamine synthetase-glutamate synthase (GS-GOGAT) pathway.

Steps of Ammonia Assimilation

- **Glutamine Synthetase (GS):** It catalyzes the ATP-dependent conversion of ammonia and glutamate into glutamine, a key nitrogen carrier in plants.
- **Glutamate Synthase (GOGAT):** It uses glutamine and 2-oxoglutarate as substrates to produce two molecules of glutamate, which serve as building blocks for the synthesis of other amino acids.

This assimilation process is tightly regulated to ensure that nitrogen is efficiently incorporated into biomolecules and is available for the plant's growth and development.

Genetic Regulation of Nitrogen Fixation

The process of biological nitrogen fixation is controlled by a complex network of genes in both the legume and the rhizobium. These genes regulate various aspects of the symbiosis, from nodule formation to nitrogenase activity, ensuring that nitrogen fixation occurs efficiently and in response to environmental cues.

Nodulation Genes (Nod Genes): Nodulation genes in rhizobia are responsible for the production of Nod factors, the signalling molecules that initiate the symbiotic interaction with the legume host. These genes are often located on large plasmids or symbiosis islands within the rhizobial genome and are highly conserved among different rhizobial species.

Key Nodulation Genes

- ***NodD*:** A transcriptional activator that binds to flavonoids secreted by the legume roots. The activation of NodD initiates the expression of other nod genes, leading to the synthesis of Nod factors.
- ***NodA, NodB, NodC*:** These genes encode enzymes involved in the synthesis of the core structure of Nod factors, which are modified by additional genes to create species-specific variations in the Nod factors.
- ***NodE and NodF*:** These genes are involved in the addition of specific fatty acids to the Nod factors, which can influence host specificity and the effectiveness of nodulation.

***Nif* Genes (Nitrogen Fixation Genes):** The nif genes encode the components of the nitrogenase enzyme complex as well as proteins involved in the electron transport chain and energy metabolism necessary for nitrogen fixation. These genes are often co-located in operons, allowing for coordinated expression in response to environmental conditions.

Key Nitrogen Fixation Genes

- ***NifH*:** Encodes the Fe protein of nitrogenase, responsible for electron transfer to the MoFe protein.
- ***NifD* and *NifK*:** Encode the subunits of the MoFe protein, which catalyzes the reduction of nitrogen to ammonia.
- ***NifL* and *NifA*:** Regulatory proteins that respond to oxygen and other environmental signals to control the expression of nif genes.

The expression of nif genes is tightly regulated to ensure that nitrogen fixation occurs only when conditions are favorable, such as low oxygen levels and the presence of a suitable carbon source.

Plant Genes Involved in Nodulation and Nitrogen Fixation: In addition to the rhizobial genes, several plant genes are involved in the establishment and maintenance of the symbiotic relationship. These include genes encoding receptor kinases that recognize Nod factors, transcription factors that regulate nodule development, and enzymes involved in the biosynthesis of leghaemoglobin and other nodule-specific proteins.

Key Plant Genes

- ***NFR1* and *NFR5*:** Receptor kinases in legumes that bind to Nod factors and initiate the signaling cascade leading to nodule formation.
- ***NIN* (Nodule Inception):** A transcription factor that plays a central role in the early stages of nodule development.
- ***Enod40*:** An early nodulin gene involved in the differentiation of root cortical cells into nodule cells.

These genes are subject to complex regulatory networks that integrate signals from the environment, the rhizobia, and the plant's own metabolic status to optimize nitrogen fixation.

Environmental Factors Influencing Nitrogen Fixation

Biological nitrogen fixation is highly sensitive to environmental conditions, and several factors can influence the efficiency of this process in grain legumes.

Soil pH: Soil pH is a critical factor affecting the survival and activity of rhizobia as well as the availability of nutrients required for nodulation and nitrogen fixation. Most rhizobia prefer a slightly acidic to neutral pH (6.0-7.0). In acidic soils (pH < 5.5), the availability of essential nutrients such as calcium and molybdenum decreases, which can impair nodulation and nitrogenase activity. Conversely, alkaline soils (pH > 8.0) can reduce the solubility of phosphorus and micronutrients, further limiting nitrogen fixation.

Strategies to manage soil pH, such as liming acidic soils or using acid-tolerant rhizobial strains, can help enhance nitrogen fixation in challenging environments. The bacteria most often involved in BNF in forages, those in the genus *Rhizobium*, are often negatively affected by soil pH values less than 6.0. If soil pH is less than 6.0, and where it is economical to do so, limestone may be added to increase soil pH and thus increase BNF. There are some exceptions to this general rule. For example the symbiosis between red clover and its microsymbiont *Rhizobium leguminosarum* biovar *trifolii* appears to function well at pH values less than 6.0 (Ranjan *et al.*, 2022).

Soil Nutrients: The availability of key nutrients, including phosphorus, potassium, and sulfur, is essential for both plant growth and nitrogen fixation. Phosphorus is particularly important for energy metabolism and the synthesis of nucleic acids and phospholipids. Potassium is involved in enzyme activation and osmoregulation, while sulfur is a component of certain amino acids and coenzymes.

In nitrogen-deficient soils, the presence of adequate phosphorus and potassium can enhance nodulation and nitrogen fixation. However, high levels of available nitrogen (e.g., from fertilizers) can inhibit nodulation and nitrogen fixation, as the plant may prioritize direct nitrogen uptake over symbiotic fixation.

Temperature: Temperature influences the growth and activity of both the host plant and the rhizobia. Most rhizobia species have an optimal temperature range for growth and nitrogen fixation, typically between 20°C and 30°C. Temperatures outside this range can reduce the efficiency of nodulation and nitrogen fixation. Temperature has a marked influence on survival and persistence of rhizobial strains in soils. For example, cowpea rhizobia from the hot dry Sahel-savannah of West Africa grow at 37°C, and more than 90% of the strains isolated from this region grew well to 40°C (Werner and Newton, 2005). The influence of temperature on rhizobia appears to be both strain and soil dependent. For example, Bradyrhizobium sp. was less susceptible than R. leguminosarum bv. trifolii to high soil temperatures, but addition of montmorillonite and illite remediated this problem in sandy soils. Soil temperature also greatly influences competition for nodulation (Triplett and Sadowsky, 1992). This effect may, in part, be due to a temperature-induced delay in nodulation or the restriction of nodules to the sub-surface region. Aranjuelo *et al.* (2007) evaluated effect of elevated temperature and water availability on CO_2 exchange and nitrogen fixation of nodulated alfalfa plants.

High temperatures can lead to the denaturation of proteins, including nitrogenase, and can disrupt membrane integrity in both the plant and the bacteria. Low temperatures can slow down metabolic processes, reducing the availability of ATP and reducing power required for nitrogen fixation.

Water Availability: Water availability is a major factor influencing nitrogen fixation, particularly in rain-fed agricultural systems. Drought stress can reduce root growth, limit the movement of rhizobia to the root zone, and decrease the rate of photosynthesis, leading to lower carbohydrate availability for the bacteroids. Soil water is related to soil pore space, and soils containing larger pores and pore spaces retain less water. Thus, soil aggregates having smaller internal pore spaces are more favorable environments for the growth of rhizobia and most soil microbes (Turco and Sadowsky, 1995). Soil-water content also directly influences the growth of rhizosphere micro-organisms, like rhizobia, by decreasing water activity below critical tolerance limits and indirectly by altering plant growth, root architecture, and exudations. In contrast, waterlogging can create anaerobic conditions in the soil, which can inhibit the respiration of rhizobia and reduce nitrogen fixation. Water management practices, such as irrigation and drainage, can help maintain optimal soil moisture levels for nitrogen fixation.

Salinity: Soil salinity is an increasing concern in many agricultural regions, particularly in arid and semi-arid areas. High salinity can create osmotic stress, reducing water uptake by the plant and leading to ionic toxicity, particularly from sodium (Na^+) and chloride (Cl^-) ions. Salinity can negatively impact rhizobia survival, root hair formation, nodulation, and nitrogen fixation. Developing salt-tolerant legume varieties and rhizobia strains is a key area of research to enhance nitrogen fixation in saline environments.

Advances in Enhancing Nitrogen Fixation in Grain Legumes

Significant progress has been made in understanding the molecular and environmental factors that influence biological nitrogen fixation. This knowledge has led to the development of new strategies to enhance nitrogen fixation in grain legumes, with the goal of improving crop productivity and sustainability.

Genetic Engineering: Genetic engineering offers the potential to enhance nitrogen fixation by introducing or overexpressing key genes involved in nodulation and nitrogen fixation. For example, engineering rhizobia with enhanced nitrogenase activity or greater tolerance to environmental stresses could lead to more efficient nitrogen fixation. In plants, genetic engineering can be used to introduce or modify genes involved in nodule formation, signaling pathways, and nitrogen assimilation. For example, overexpressing genes involved in the biosynthesis of leghemoglobin or other nodule-specific proteins could enhance the efficiency of nitrogen fixation.

Breeding for Enhanced Symbiosis: Traditional breeding and modern genomic approaches are being used to develop legume varieties with enhanced symbiotic efficiency. This includes selecting for traits such as increased root exudation of flavonoids, improved nodulation efficiency, and greater tolerance to environmental stresses. Marker-assisted selection and genomic selection are being used to accelerate the breeding process by identifying and selecting for key genetic loci associated with enhanced nitrogen fixation. These approaches can help develop legume varieties that are better adapted to specific environments and that can maintain high levels of nitrogen fixation under suboptimal conditions.

Inoculants and Biofertilizers: The use of rhizobial inoculants is a well-established practice to enhance nitrogen fixation, particularly in soils with low native rhizobia populations or in new cultivation areas. Advances in inoculant technology, including the development of more effective strains, improved formulations, and better delivery systems, have led to more consistent and reliable results. Inoculants can be tailored to specific legume species and environmental conditions, ensuring that the most effective rhizobia are present in the root zone. The development of biofertilizers that combine rhizobia with other beneficial microorganisms, such as mycorrhizal fungi or phosphate-solubilizing bacteria, offers the potential to further enhance nitrogen fixation and overall plant growth.

Challenges and Future Perspectives

Despite the significant advances in understanding and enhancing biological nitrogen fixation, several challenges remain. One of the key challenges is the variability in nitrogen fixation efficiency, which can be influenced by a range of environmental factors, including soil type, climate, and management practices. Another challenge is the potential for competition between native and introduced rhizobia, which can affect the establishment and effectiveness of inoculants. Understanding the complex interactions between different microbial communities in the rhizosphere is essential for optimizing nitrogen fixation.

Future research should focus on developing resilient legume-rhizobia partnerships that can thrive under diverse and changing environmental conditions. This includes exploring the potential for introducing nitrogen fixation into non-leguminous crops, such as cereals, through genetic engineering or microbial symbiosis. Advances in systems biology, including genomics, transcriptomics, proteomics, and metabolomics, offer new tools for dissecting the complex regulatory networks underlying nitrogen fixation. Integrating these approaches with traditional breeding and biotechnology holds promise for developing legume varieties with enhanced nitrogen fixation capabilities.

Conclusion

Biological nitrogen fixation in grain legumes is a cornerstone of sustainable agriculture, providing a natural and eco-friendly alternative to synthetic nitrogen fertilizers. The intricate mechanisms underlying this process, from the initial plant-microbe interaction to the assimilation of fixed nitrogen, highlight the complexity and efficiency of this symbiotic relationship. Environment condition such as temperature, pH, nutrient availability and soil condition had a significant difference in nitrogen fixation

As global food demands continue to rise, leveraging and enhancing biological nitrogen fixation in grain legumes will be crucial for improving crop productivity, reducing environmental impacts, and ensuring long-term agricultural sustainability. Ongoing research and technological innovations will play a key role in achieving these goals, contributing to a more resilient and sustainable agricultural system.

References

Aranjuelo, I., Irigoyen, J., & Sanchez-Dıaz, M. (2007). Effect of elevated temperature and water availability on CO_2 exchange and nitrogen fixation of nodulated alfalfa plants. Environmental and Experimental Botany, 59, 99-108.

Beijerinck, M. W. (1901). Uber oligonitrophile Mikroben. Zbl. Backt, 7, 561-582.

Berrada, H., & Fikri-Benbrahim, K. (2014). Taxonomy of the rhizobia: current perspectives. British Microbiology Research Journal, 4(6), 616.

Chaudhary, A., Gustafson, D., & Mathys, A. (2018). Multi-indicator sustainability assessment of global food systems. Nature Communications, 9, 848.

FAO (2017). World Fertilizer Trends and Outlook to 2020. (Available online at: http://www.fao.org/3/a-i6895e.pdf)

Gilbert, N. (2012). One-third of our greenhouse gas emissions come from agriculture. Nature, 10(38), 11708.

Glendining, M. J., Dailey, A. G., Williams, A. G., Evert, F. K. V., Goulding, K. W. T., & Goulding, A. P. (2009). Is it possible to increase the sustainability of arable and ruminant agriculture by reducing inputs. Agricultural Systems, 99, 117-125.

Hubbell, D. H., & Kidder, G. (2009). Biological Nitrogen Fixation. University of Florida IFAS Extension Publication, 16, 1-4.

Ladha, J. K., Tirol-Padre, A., Reddy, C. K., Cassman, K. G., Verma, S., & Powlson, D.S. (2016). Global nitrogen budgets in cereals: A 50-year assessment for maize, rice, and wheat production systems. Scientific Reports, 6, 19355.

Leinonen, I., Iannetta, M. P., Rees, R. M., Russell, W., Watson, C., & Barnes, A. P. (2019). Lysine supply is a critical factor in achieving sustainable global protein economy. Frontiers in Sustainable Food System, 3, 27.

MacWilliam, S., Wismer, M., & Kulshreshtha, S. (2014). Life cycle and economic assessment of western canadian pulse systems: the inclusion of pulses in crop rotations. Agricultural Systems, 123, 43-53.

Postgate, J. R. (1982). The Fundamentals of Nitrogen Fixation. New York: Cambridge University Press.

Ranjan, S., Kumar, S., Dutta, S. K., Sow, S., Kumar, S., & Sushant. (2023). Long-term organic amendment application improves soil fertility status, nutrient accumulation and crop productivity under rice-wheat cropping system. Communications in Soil Science and Plant Analysis, 54(18), 2579-2589.

Ranjan, S., Sow, S., Kumar, S., Ghosh, M., Roy, D. K., Dutta, S. K., & Pramanick, B. (2022). Enhancing Nutrient Use Efficiency and Productivity of Cereals through Site Specific Nutrient Management. Journal of Cereal Research, 14(3), 229-242.

Raun, W. R., & Johnson, G. V. (1999). Improving nitrogen use efficiency for cereal production. Agronomy Journal, 91, 357-363.

Smykal, P., Coyne, C. J., Ambrose, M. J., Maxted, N., Schaefer, H., & Blair, M. W. (2015). Legume crops phylogeny and genetic diversity for science and breeding. Crit. Rev. Plant Science, 34, 43-104.

Sow, S., Singh, G., Ghosh, M., Dutta, S. K., Mandal, N., Kumar, S., & Ranjan, S. (2023). Nitrogen-management strategy through leaf-colour chart and SPAD meter for optimizing the productivity in irrigated wheat (Triticum aestivum). Indian Journal of Agronomy, 68(2), 219-222.

Triplett, E.W., & Sadowsky, M.J. (1992). Genetics of competition for nodulation. Annals Review Microbiology, 46, 399-428.

Turco, R.F., & Sadowsky, M.J. (1995). Understanding the microflora of bioremediation. In H.D. Skipper and R. F. Turco (Eds.), Bioremediation: Science and Applications. Soil Science (Special Publication). Madison, WI: Soil Sci. Soc. American Journal, 43, 87-103.

Udvardi, M., & Poole, P.S. (2013). Transport and metabolism in legume-rhizobia symbioses. Annual Review of Plant Biology, 64, 781-805.

Vance, C. P. (2001). Symbiotic nitrogen fixation and phosphorus acquisition. Plant nutrition in a world of declining renewable resources. Plant Physiology, 127(2), 390-397.

Werner, D. & Newton, W.E. (2005). Nitrogen fixation in agriculture, forestry, ecology, and the environment. Springer, pp 32-36.

3

Biotic Stress Mitigation Strategies in Grain Legumes

Debasish Rout[1], Manisha Rout[2*] and Pritam Ghosh[2]

[1]*Synergy Institute of Engineering & Technology, Dhenkanal, India*

[2]*Department of Agronomy, Institute of Agriculture, Visva-Bharati University West Bengal, India*

**Corresponding author's email: manisha.rout55@gmail.com*

Abstract

Biotic stress in plants refers to the adverse impact of living organisms, such as pathogens, pests, and nematodes, on plant growth, development, and productivity. These biotic agents can infect, infest, or parasitize plants, triggering a range of physiological and biochemical responses that compromise plant health. Biotic stress can manifest in various forms, including fungal diseases, bacterial blights, viral infections, insect infestations, and nematode attacks, leading to reduced crop yields, lower quality produce, and significant economic losses. Understanding biotic stress is crucial for developing effective management strategies to protect plants and ensure food security, as it poses a significant threat to agricultural productivity and sustainability. To enhance the productivity of grain legumes, a critical food source for millions, mitigating biotic stress is essential. By adopting effective strategies to manage biotic stressors like diseases, pests, and nematodes, legume yields can be significantly increased. This can be achieved through the development and deployment of resistant cultivars, integrated pest management (IPM) practices, cultural methods like crop rotation and sanitation, and biological control methods. Additionally, rhizobia and mycorrhizal inoculation can enhance plant defense and nutrient uptake, while precision agriculture techniques facilitate early detection and targeted management of biotic stress. By implementing these approaches, legume productivity can be boosted, ensuring food security, improving livelihoods, and meeting the growing demand for this vital crop.

Keywords: *Biotic stress, Integrated pest management, Legume productivity, Sustainability*

Introduction

Legumes are indeed a ubiquitous crop, grown in over 120 countries across all continents, showcasing their adaptability and importance in diverse agricultural systems. From the soybean fields of North America to the chickpea farms of Asia, and from the cowpea cultivation in Africa to the lentil production in Europe, legumes play a vital role in food security, nutrition, and sustainable agriculture worldwide. Their global presence underscores the need for continued research, development, and innovation to enhance legume productivity, resilience, and sustainability, ensuring their benefits are shared by farmers, consumers, and the environment alike (Ma *et al.*, 2024, El-Hassan *et al.*, 2013).

Grain legumes hold immense potential for transforming the agricultural scenario and human health, but their growth is hindered by various challenges. As the availability of agricultural land is decreasing per person and the human population is rising, the production of foodgrains per unit of land needs to be increases and the wastage of food grains need to be kept to the minimum. To unlock this to the best, it is crucial to bridge the yield gap between achievable yield potential and achieved yield of grain legumes through innovative research, enhancement of water use efficiency, and improvement of soil health. Furthermore, developing climate-resilient varieties, reducing post-harvest losses, and creating market opportunities can boost grain legume production and consumption. By addressing these critical issues, we can harness the power of grain legumes to enhance food security, improve nutrition, and support sustainable agriculture, ultimately contributing to a food-secure and prosperous future (Abobatta *et al.*, 2021).

Climate change is significantly affecting grain legume production worldwide. Rising temperatures and altered precipitation patterns are leading to reduced yields, lower quality produce, and increased susceptibility to pests and diseases. Global warming, rising carbon dioxide concentration, pollution, industralisation, farm mechanization in agriculture are all together a very difficult challenge in world agriculture. The book, "*Handbook of plant and crop stress*" by Pessarakli (2019), has given a clear picture about various stresses in plants which is summarized below with regards to grain legumes.

Types of Stress in Grain Legumes

Grain legumes face various types of stress that impact their growth, productivity, and quality. These stresses can be broadly categorized into:

1. **Biotic Stress**: Caused by living organisms like:
    - Insect Pests (insects, mites, and nematodes)

- Diseases (fungal, bacterial, and viral)
- Weeds (plants out of place competing for resources)

2. **Abiotic Stress:** Caused by non-living factors like:
   - Drought (water scarcity)
   - Temperature extremes (heat, cold, and frost)
   - Salinity (soil salt content)
   - Nutrient deficiencies (soil nutrient imbalance)
   - Waterlogging (excess water)
3. **Physiological Stress:** Caused by internal factors like:
   - Oxidative stress (cellular damage)
   - Hormonal imbalance
   - Senescence (aging)
4. **Environmental Stress:** Caused by external factors like:
   - Climate change (rising temperatures, changing precipitation patterns)
   - Air pollution (gaseous and particulate pollutants)
   - Soil degradation (erosion, compaction, and nutrient depletion)

Understanding these types of stress is crucial for developing effective management strategies to enhance grain legume resilience, productivity, and sustainability. Plants are constantly exposed to various biotic stressors, including fungi, bacteria, insects, viruses, nematodes, and weeds. Effective and sustainable management of these pathogens and insect pests is essential for all agricultural production systems. However, pathogens continue to develop resistance to the currently available resistance genes and insecticides. Climate change may further affect plant-insect dynamics, altering host plant resistance, the role of natural enemies, and insect interactions, which in turn can influence insect physiology and behavior. Increased levels of carbon dioxide may intensify infestations by certain pests, such as lepidopteran pod-borers and coleopteran defoliators. Additionally, secondary pest infestations have been rising rapidly, leading to significant losses in crop production.

Biotic Stresses and their Effects On Major Legume Crops

Legume crops face a wide range of biotic stresses that significantly affect their productivity and pose a threat to global food security. Insect herbivory and viral infections reduce leaf area and disrupt photosynthesis, while stress-inducing bacteria can endanger plant survival by interfering with nutrient delivery. Temperature fluctuations also play a critical role by influencing pest activity and reproduction, contributing to substantial annual yield losses due to pests and diseases in legume crops. In some regions, such as Asia and Africa, legume crops can suffer catastrophic losses of up to 100% when conditions favor biotic stress. This section of the article examines the impact of various biotic stresses on important legume crops, highlighting the severe damage and yield losses caused by these stressors.

Soybean crops worldwide face a similar challenge, with over 20 diseases affecting major soybean-producing countries. Root-knot nematodes alone cause a substantial reduction in soybean productivity, while a variety of insect pests-including whiteflies, jassids, aphids, soybean leaf folders, green semiloopers, leaf webbers, and pod borers-further contribute to a 30-40% decline in yield.

Chickpea yields, particularly in India, also fall short of their potential, averaging around 1.0-1.2 tonnes per hectare compared to the predicted potential of over 2 tonnes per hectare (Hesler *et al.*, 2018). Mung bean, despite a high yield potential of 2.5-3.0 tonnes per hectare, often achieves only about 0.5 tonnes per hectare due to a combination of abiotic and biotic constraints, inadequate crop protection, and limited access to improved seeds. One major viral disease, Mungbean Yellow Mosaic Disease (MYMD), has caused significant economic losses, reducing yields in India by as much as 85%. Additionally, other fungal diseases, such as *Rhizoctonia* root rot, anthracnose, powdery mildew, and *Cercospora* leaf spot, further reduce mung bean yields. Emerging threats like halo blight have caused severe output losses and crop failure in Chinese mung bean farms, while countries like India and Iran have also suffered significant yield reductions due to bacterial leaf spot caused by *Xanthomonas phaseoli* (Kumar and Kirti, 2018).

Mitigation Approaches Towards Biotic Stress Management

Genetic resistance is the most effective strategy for mitigating biotic stress in legumes. However, due to the variable nature of resistance, the diversity of pathogen strains, and the wide range of host plants, a single approach is not sufficient. Additionally, the narrow genetic base and limited screening methods for genetic resources pose challenges in achieving durable resistance to diseases and insects in grain legumes.

To fully exploit the genetic potential of legumes for resistance and to enhance the availability of genomic resources, integrating pre-breeding approaches with genomic breeding methods could be beneficial. This integration can help in developing high-yielding cultivars with resistance to biotic stresses, thereby boosting legume productivity in key growing areas. Advanced biotechnological tools, such as whole genome sequencing, transcriptomics, proteomics, advanced mutagenesis methods, genome editing, and genome-wide association studies, will be crucial in understanding the genetic interactions, proteins, and metabolites involved in conferring resistance to biotic stresses.

Mitigation Strategies Against *Sclerotinia* Diseases in Legume Crops

Legumes, a crucial food source for humans and animals, are under threat from Sclerotinia diseases caused by the fungal pathogen *Sclerotinia sclerotiorum*. This devastating disease affects over 600 plant species, including legumes, compromising global food security. Antwi-Boasiako *et al.* (2022) in their long termed study have found that, among the various management strategies investigated, genetic control poses a significant challenge due to the scarcity of resistant traits within existing germplasm (Fig. 3.1). However, biological agents have emerged as a promising solution, demonstrating encouraging results in combating *Sclerotinia* diseases in grain legume crops. It is known for causing disease among legumes such as groundnut (*Arachis hypogaea* L.), soybean (*Glycine max* L.), common bean (*Phaseolus vulgaris* L.), faba bean (*Vicia faba* L.), alfalfa (*Medicago sativa* L.), red clover (*Trifolium pratense* L.), chickpea (*Cicer arietinum* L.), and lupin (*Lupinus albus* L.)

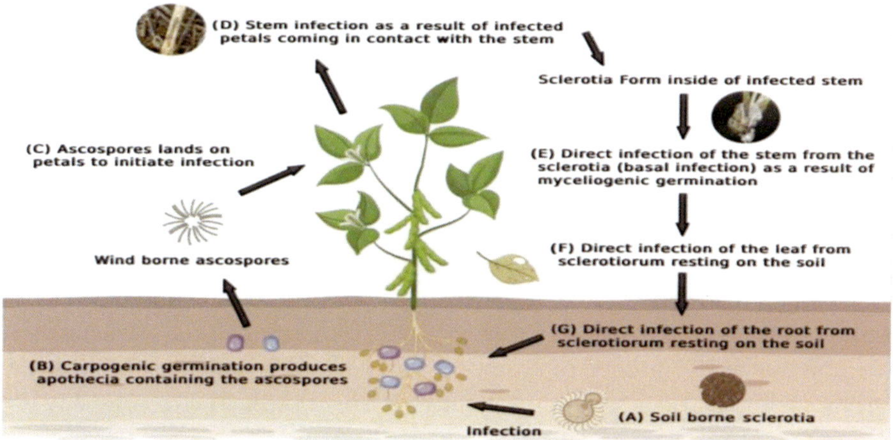

Fig. 3.1: Infection process of S. *sclerotiorum*

- **Biological Control:** Biological control offers an eco-friendly alternative to chemical pesticides for managing S. sclerotiorum. Various microorganisms, including bacteria, fungi, nematodes, viruses, and insects, can be used as biological control agents (BCAs). BCAs are crucial for sustainable agriculture and have been shown to effectively control *S. sclerotiorum* in legume crops.

Coniothyrium minitans is good in controlling *S. sclerotiorum* in crop of soybean. It is usually available in the formulation of Contains WG. In soybean plants, the application of *Coniothyrium minitans* effectively slows down apothecia and sclerotia formation by 81% and 50%, respectively (Selin *et al.*, 2010).

- **Plant Growth Promoters and Biocontrol Agents:** Certain Bacillus species have been identified as plant growth-promoting rhizobacteria (PGPR) and are used as biological control agents (BCAs). These beneficial bacteria not only enhance the nutritional properties of crops like *Amaranthus hypochondriacus* Linn but also exhibit antagonistic effects against *Sclerotinia sclerotiorum* in various legume crops. Sabaté *et al.* (2018), have demonstrated the potential of *Bacillus* species in:
- Promoting plant growth and nutrition
- Controlling *S. sclerotiorum* incidence in legume crops
- Offering a sustainable and eco-friendly alternative to chemical pesticides

The use of Bacillus species as BCAs and PGPR highlights their significance in agriculture, particularly in managing plant diseases and improving crop productivity.

- **Genetic Improvement of Host Resistance:** Developing resistant legume varieties is essential to combat S. sclerotiorum. While some resistant genes have been identified, breeding programs have had limited success, and no commercial varieties are available for most legume crops. Marker-assisted breeding and host-induced gene silencing show promise in managing the disease (Robison *et al.*, 2018).

Molecular markers have helped identify partial resistance in legumes like common bean and soybean. Quantitative trait loci (QTL) linked to *S. sclerotiorum* resistance have been mapped in several legume crops, but breeding programs have had limited success. No commercial varieties are available for most legume crops, except for groundnut, which has a resistant variety saving the US $5 million annually (Bakhshi *et al.*, 2020).

Recent studies have identified resistant genes in soybean and demonstrated the potential of host-induced gene silencing in managing the disease. Genomic regions linked to resistance have been found in common bean, offering a foundation for marker-assisted breeding. Developing resistant cultivars will reduce the reliance on fungicides.

- **Chemical Control:** Fungicides are widely used to manage *S. sclerotiorum*, but their application poses environmental and health risks. Rotating fungicides with diverse modes of action and applying them at critical crop stages can improve efficacy. However, chemical control is considered uneconomical by farmers and can have negative ecological consequences (Khangura and Beard, 2015)

Mitigating Insect Pests in Grain Legumes

The overuse of chemical pesticides has led to the emergence of pesticide-resistant pests, prompting a shift towards eco-friendly alternatives. Biopesticides, or BCAs (Biological Control Agents), have gained popularity as a safe and environmentally friendly option for pest management. Various BCAs, such as *Trichoderma* spp., *Pseudomonas* spp., and *Bacillus* spp., have been identified and used to control pests and diseases. The shift towards biopesticides is a positive step towards sustainable agriculture, but more efforts are needed to increase adoption rates and develop innovative BCA technologies.

Biological Control Agents (BCAs) and their Formulations for Pest Management in Legumes

BCAs have been used to manage pests in legumes by incorporating BCA-colonized natural substrates into the rhizosphere (Singh and Prasad, 2014). Examples include:

- Bacterial endophytes like *Bacillus, Paenibacillus*, and *Pseudomonas* showing antifungal activity against major pathogens.
- *Trichoderma* species effective against wilt and root rot diseases.
- Fungi like *Metarhizium anisopliae*, *Beauvaria bassiana*, and *Paecilomyces lilacinus* employed as biopesticides.
- *Bacillus thuringiensis* (Bt) and Nuclear Polyhedrosis Viruses (NPVs) used as biopesticides.

Discovery of QTLs for Disease Resistance

Quantitative Trait Loci (QTL) mapping is an essential method used to link plant disease resistance loci with specific resistance traits. This approach allows for the detailed analysis of individual resistance loci, the specificity of resistance genes, and their interactions with each other, as well as with the plant's developmental stage and the environment. Research on the identification of QTLs for resistance to soil-borne diseases has been limited, particularly for crops such as common beans and cowpeas. In cowpeas, about eight QTLs associated with resistance have been identified, with one significant QTL, Mac-2, accounting for over 40% of the resistance diversity (Böttger *et al.*, 2018). A marker linked to the pectin esterase inhibitor (PEI)-encoding gene 1 0853 has also been identified; since pectins play a vital role in pathogen defense, the PEI gene has been recognized as a phaseolina-resistant gene for *M. cowpea*. Genome-Wide Association Studies (GWAS) have recently gained importance for identifying resistance QTLs, though knowledge of QTLs related to soil-borne diseases remains limited.

Genetic Engineering for Crop Insect/Pest Resistance

Genetic engineering techniques offer promising strategies for developing insect and pest resistance in legumes. Cry proteins, which are insecticidal crystal toxins derived from soil-dwelling gram-positive bacteria, can be expressed alone or in combination with other genes, such as vegetative insecticidal proteins (VIP), proteinase inhibitors, and lectins, to enhance resistance to insect pests. Advanced biotechnological approaches, including RNA interference (RNAi) and CRISPR/Cas9, are also viable for long-term insect pest control. Figure 2 illustrates various genetic engineering approaches used to develop insect-resistant legume varieties.

Cry toxins, also known as insecticidal crystal toxins or crystalline inclusions, are proteins that harm insects from several families, including Coleoptera, Diptera, Lepidoptera, Hymenoptera, and Nematoda, while remaining safe for humans, animals, and natural insect predators.

Transgenic Legumes

Transgenic approaches have shown potential in conferring resistance to various pests in legumes. For example, the overexpression of the agglutinin gene in chickpeas has been found to significantly reduce the survival and fertility of cowpea aphids (*Aphis craccivora*), which are known to cause considerable damage to chickpea crops. Transgenic pigeon pea expressing *Glycine max* trypsin inhibitor and Cry1Ab has also been developed, though it does not effectively inhibit *Helicoverpa armigera*. Other studies have demonstrated that transgenic chickpeas expressing inhibitors of cowpea trypsin and alpha-amylase can provide resistance to bruchids. Similarly, resistance to pea weevil has been achieved in transgenic pea by expressing an inhibitor of amylase.

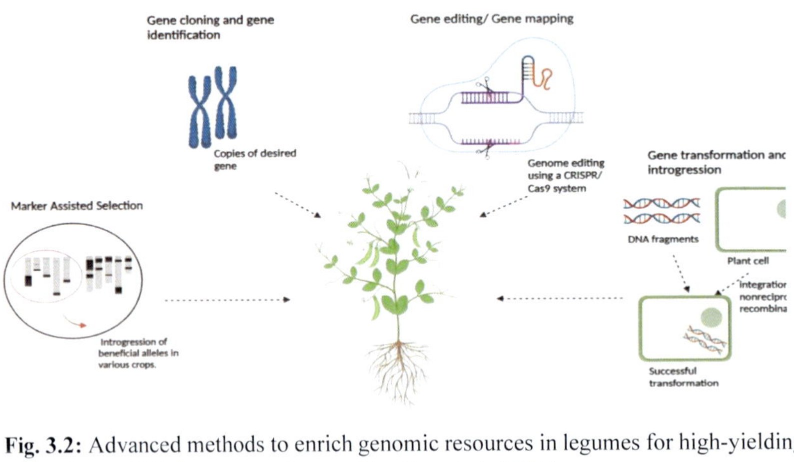

Fig. 3.2: Advanced methods to enrich genomic resources in legumes for high-yielding cultivars with enhanced biotic stress resistance (Sandhu *et al.*, 2023)

These genetic and biotechnological interventions are vital tools in developing legume crops with enhanced resistance to diseases and pests, contributing to more sustainable and productive agricultural systems.

Weed Management in Grain Legumes

Legumes with their quick growth and development, are effective in weed suppression owing to their ground coverage with broad leaves. They are weed suppressors in many cash crops like sweetcorn, and other cereals. Shading effect of legumes on the tiny weeds gives them competitive advantage for natural resources (Gogoi *et al.,* 2018). Thus, legumes should be bred for faster initial growth and development, sowing time can be adjusted by weed shift and weed occurrence data, micro-irrigation can aid in wetting the root rhizosphere, and other good agronomic practices need to be taken holistically for better competitive advantage and stress mitigation in grain legumes.

Weed management is a critical aspect of grain legume production, as weeds can significantly reduce yields, lower crop quality, and increase production costs. Effective weed management strategies are essential to minimize these impacts and ensure sustainable grain legume production. Weeds in grain legumes can be managed through a combination of cultural, mechanical, and chemical methods. Cultural practices such as crop rotation, sanitation, and proper fertilization can help prevent weed growth. Mechanical methods like hand-weeding, hoeing, and mulching can also be effective. Chemical control using herbicides can be used as a last resort, but it requires careful selection and application to avoid harming the crop or the environment. Integrated weed management (IWM) approaches that combine multiple methods are often the most effective and sustainable way to manage weeds in grain legumes. IWM considers the entire production system and uses a range of techniques to minimize weed growth and impact. By adopting IWM strategies, grain legume producers can reduce weed pressure, improve crop yields, and promote environmental sustainability (Roy *et al*, 2023).

Conclusion

Biotic stress mitigation strategies in grain legumes are crucial for ensuring food security, sustainable agriculture, and environmental conservation. While significant progress has been made, there remains a need for integrated solutions that combine multiple strategies, prioritize climate-resilient and sustainable practices, and advance our understanding of plant-microbe interactions. Furthermore, developing cost-effective and accessible technologies for smallholder farmers is essential. By adopting a holistic approach to biotic stress mitigation, we can enhance the resilience of grain legume crops, improve yields, and contribute to a more sustainable and food-secure future. Through the adoption of integrated pest management (IPM), integrated disease management (IDM), and integrated weed management (IWM) approaches, legume producers can minimize the use of chemical pesticides and fungicides, reduce environmental pollution, and promote ecosystem services.

Furthermore, advances in genetic improvement, biotechnology, and precision agriculture offer exciting opportunities for developing climate-resilient and stress-tolerant legume varieties, optimizing crop management practices, and enhancing legume productivity and sustainability. Ultimately, effective biotic stress management in legumes requires a collaborative effort from researchers, extension agents, farmers, and policymakers to develop and disseminate innovative technologies and strategies, promote sustainable agricultural practices, and ensure global food security.

References

Abobatta, W. F., El-Hashash, E. F., & Hegab, R. H. (2021). Challenges and opportunities for the global cultivation and adaptation of legumes. Journal of Applied Biotechnology & Bioengineering, 8(5), 160–172.

Antwi-Boasiako, A., Wang, Y., Dapaah, H. K., & Zhao, T. (2022). Mitigating against Sclerotinia diseases in legume crops: A comprehensive review. Agronomy, 12(12), 3140. https://doi.org/10.3390/agronomy12123140

Bakhshi, S., Zadeh, H. R. Z., & Atghia, O. (2020). Molecular identification of bean genotypes for partial resistance to Sclerotinia sclerotiorum based on SCAR markers. Journal of Plant Pathology, 102, 369–375.

Böttger, A., Vothknecht, U., Bolle, C., & Wolf, A. (2018). Plant secondary metabolites and their general function in plants. In Lessons on Caffeine, Cannabis & Co: Plant-Derived Drugs and Their Interaction with Human Receptors (pp. 3–17). Springer Nature.

El-Hassan, S. A., Gowen, S. R., & Pembrok, B. (2013). Use of Trichoderma hamatum for bio-control of lentil wilts disease. Journal of Plant Protection Research, 53(1), 12–17.

Gogoi, N., Baruah, K. K., & Meena, R. S. (2018). Grain legumes: Impact on soil health and agroecosystem. In Legumes for Soil Health and Sustainable Management, pp. 511–539.

Hesler, L. S., Allen, K. C., Luttrell, R. G., Sappington, T. W., & Papiernik, S. K. (2018). Early-season pests of soybean in the United States and factors that affect their risk of infestation. Journal of Integrated Pest Management, 9(1), 1–15.

Khangura, R., & Beard, C. (2015). Managing Sclerotinia stem rot in canola. Department of Agriculture and Food, Australian Government. Available online: https://www.agric.wa.gov.au/canola/managing-sclerotinia-stem-rot-canola (accessed on August 8, 2024).

Kumar, D., & Kirti, P. B. (2015). Transcriptomic and proteomic analyses of resistant host responses in Arachis diogoi challenged with late leaf spot pathogen, Phaeoisariopsis personata. PLOS ONE, 10(2), e0117559. https://doi.org/10.1371/journal.pone.0117559

Ma, L., Ma, S., Chen, G., Lu, X., Chai, Q., & Li, S. (2024). Mechanisms and mitigation strategies for the occurrence of continuous cropping obstacles of legumes in China. Agronomy, 14(1), 104. https://doi.org/10.3390/agronomy14010104

Pessarakli, M. (Ed.). (2019). Handbook of Plant and Crop Stress. CRC Press.

Robison, F. M., Turner, M. F., Jahn, C. E., Schwartz, H. F., Prenni, J. E., Brick, M. A., & Heuberger, A. L. (2018). Common bean varieties demonstrate differential physiological and metabolic responses to the pathogenic fungus Sclerotinia sclerotiorum. Plant, Cell & Environment, 41, 2141–2154.

Roy, D. K., Ranjan, S., & Sow, S. (2023). Weed management effect on weeds, productivity and economics of soybean. Indian Journal of Weed Science, 55(2), 228-230.

Sabaté, D. C., Brandan, C. P., Petroselli, G., Erra-Balsells, R., & Audisio, M. C. (2018). Biocontrol of Sclerotinia sclerotiorum (Lib.) de Bary on common bean by native lipopeptide-producer Bacillus strains. Microbiological Research, 211, 21–30.

Sandhu, R., Bangarwa, S. K., Attri, M., Tiwari, S., Kohli, S., Fayaz, S., & Chaudhary, N. (2023). Effects of biotic stresses and their mitigation strategies in legumes: A review. Legume Research. https://doi.org/10.18805/LR-5160.

Selin, C., Habibian, R., Poritsanos, N., Athukorala, S. N., Fernando, D., & de Kievit, T. R. (2010). Phenazines are not essential for Pseudomonas chlororaphis PA23 biocontrol of Sclerotinia sclerotiorum, but do play a role in biofilm formation. FEMS Microbiology Ecology, 71, 73–83.

Singh, A. U., & Prasad, D. (2014). Efficacy of fungal bioagents against Rotylenchulus reniformis and other antagonists on sporulation of Botrytis cinerea on dead lily leaves exposed to field conditions. Phytopathology, 85, 393–400.

4

Carbon and Nitrogen Assimilation in Grain Legumes

R. Sowmiya* and K. Kiruthika

Dept. of Agronomy, Vanavarayar Institute of Agriculture, Pollachi, Tamil Nadu, India
**Corresponding author's email: sowmiagro94@gmail.com*

Abstract

Grain legumes, members of the Fabaceae family, are essential for global nutrition and sustainable agriculture. The legumes such as soybean, chickpea, lentil, pigeon pea, black gram, green gram, faba bean and field peas are renowned for their high protein and energy content. A core aspect of their significance lies in their ability to fix atmospheric nitrogen through symbiotic relationships with Rhizobium bacteria, enhancing soil fertility and reducing the need for synthetic fertilizers. This process, coupled with their role in carbon sequestration, positions legumes as crucial components of sustainable farming systems. Grain legumes, as C_3 plants, rely on efficient carbon assimilation to convert atmospheric CO_2 into biomass, supported by high nitrogen content for optimal photosynthesis. Additionally, the symbiotic nitrogen fixation within root nodules enables legumes to transform atmospheric nitrogen into forms usable by plants, thereby contributing to soil nitrogen pools and reducing greenhouse gas emissions. By integrating legumes into crop rotations, farmers can enhance soil health, lower fossil energy inputs, and support ecological sustainability. This comprehensive analysis underscores the critical roles of carbon and nitrogen assimilation in maximizing the environmental and agronomic benefits of grain legumes.

Keywords: *Carbon assimilation, Grain legumes, Nitrogen assimilation, Nitrogen fixation, Photosynthesis*

Introduction

The grain legumes belong to the Fabaceae family, also known as Leguminosae. The word "legume" comes from the Latin *legumen*, which refers to seeds collected from pods. This group includes soybeans, chickpeas, cowpeas, pigeon peas, field peas, green grams, black grams, faba beans, and lentils. Grain legumes are primarily cultivated for their seeds, which are harvested when fully mature. They are notable for their high energy and protein content. The mature, dried seeds are consumed directly as food or feed for both humans and animals (Singh *et al.,* 2007).

Most important legumes are herbaceous, often annuals, and are grown globally. Grain legumes can be classified into two categories: (1) oilseeds, such as soybeans and groundnuts, which are grown for their protein and oil; and (2) pulses, including common beans and peas, which are primarily grown for their protein content and used as whole or split seeds in various dishes. These legumes are a key protein source for vegetarians and are an essential part of diets worldwide. They provide complex carbohydrates, soluble fibers, essential vitamins, minerals, and polyphenols like flavonoids, isoflavones, and lignans, which may contribute to their role as functional foods. Grain legumes are crucial for providing protein and energy to monogastric animals. Legumes such as faba beans, peas, and lupins can partially or entirely replace traditional animal-based protein sources, including meat, bone meal and fish meal.

Legumes have the ability to fix atmospheric nitrogen in the soil through biological processes. Typically, legumes are distinguished by their capacity to acquire nitrogen from the atmosphere with the help of nitrogen-fixing bacteria from the genus *Rhizobium* (Zahran, 1999). As a result, legume crops generally need only minimal nitrogen fertilizer. Additionally, integrating legumes into crop rotation systems can reduce the need for nitrogen fertilizer in subsequent crops (Rochester *et al.,* 1998). Grain legumes and production systems that include legumes can play a significant role in providing various benefits, aligning with sustainability principles. Legumes play a key role in carbon sequestration in soils; reduce greenhouse gas (GHG) emissions, such as carbon dioxide (CO_2) and nitrous oxide (N_2O), compared to systems that rely on mineral nitrogen fertilizers and lower the overall fossil energy inputs required in the system (Stagnari *et al.,* 2017, Ranjan *et al.*, 2022).

Carbon assimilation in legumes is crucial for their growth and productivity, as it facilitates the conversion of atmospheric carbon into plant biomass. A thorough understanding of how carbon and nitrogen are assimilated in grain legumes is vital for maximizing their advantages in agricultural systems. This knowledge enhances soil fertility and supports sustainable farming practices. This chapter delves into the essential roles of carbon and nitrogen assimilation in grain legumes and their influence on agricultural sustainability.

Photosynthesis and Carbon Assimilation in Grain Legumes

Carbon assimilation is the process by which plants and algae transform atmospheric carbon from inorganic molecules into organic matter, which serves as the foundation for biological structures and energy for cellular respiration.

Plants absorb carbon dioxide from the atmosphere through stomata in their leaves and convert it into biomass through photosynthesis (Sow *et al.*, 2024). Subsequently, they release about half of the captured carbon dioxide back into the atmosphere during respiration.

Legumes are categorized into cool season (C_3) and warm season (C_4) types, but all grain legumes are C_3 plants. Both C_4 and C_3 plants use photosynthesis to convert the 'physical energy' of photons into 'chemical energy' in the form of ATP and reduced metabolic intermediates, mainly NADPH. These compounds are then utilized in synthesizing various carbons and nitrogen assimilates particularly carbohydrates and amino acids (Foyer *et al.*, 2001). Photosynthesis is affected by factors such as solar radiation, temperature, water availability, and mineral nutrition. C_3 plants thrive within an optimal temperature range of 65-75°F and can begin growing at soil temperatures of 40-45°F. However, their efficiency decreases as temperatures rise, although they tend to produce higher protein content.

In the dark reactions, or Calvin cycle, atmospheric carbon dioxide (CO_2) enters the stroma of the chloroplast and reacts with a five-carbon molecule to form two three-carbon molecules. This process, known as C_3 photosynthesis, defines these plants as C_3 plants. The initial fixation of CO_2 is facilitated by the enzyme ribulose-bisphosphate carboxylase (RuBisCO), which is believed to have the highest biomass among enzymes globally. The two three-carbon molecules then undergo a series of reactions, each driven by different enzymes, to produce a net molecule of glucose. The formation of glucose requires high-energy compounds, particularly ATP and NADPH. As these compounds are used up in the dark reactions, they are replenished during the light reactions. Consequently, the light and dark reactions are interdependent, with each process relying on the other to continue functioning.

A report on CO_2 assimilation, leaf composition, and nitrogen supply indicates that nitrogen is essential for amino acid synthesis, which leads to the production of proteins and other cellular components. The efficient CO_2 assimilation demands considerable amounts of various chloroplast components, including light-harvesting chlorophyll-protein complexes (LHCP), elements involved in electron transport and NADP+ reduction in the thylakoids, and the CO_2-assimilating enzyme ribulose-1,5-bisphosphate carboxylase-oxygenase (Rubisco), along with other enzymes necessary for CO_2 assimilation in the stroma (Lawlor, 2002).

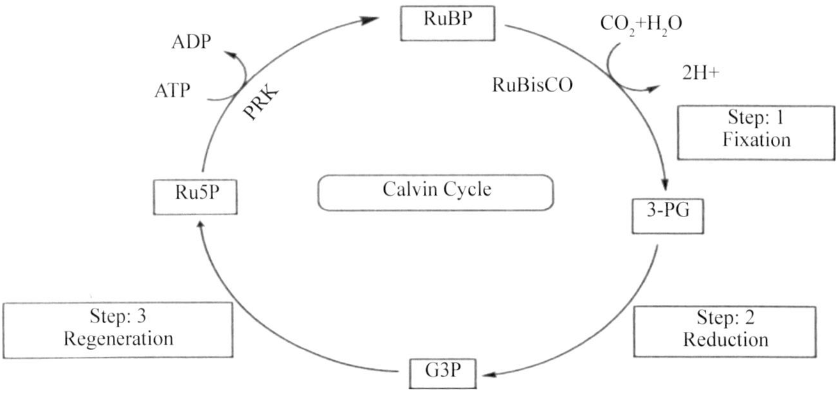

Fig. 4.1: C_3 cycle pathway in plants (Mondal *et al.*, 2015)

To efficiently capture the energy used in photosynthesis, a high concentration of chlorophyll per unit area is required, supported by proteins such as light-harvesting chlorophyll-protein complexes (LHCP) in the thylakoids, which results in a significant amount of nitrogen in the thylakoids. Rubisco, the primary protein in chloroplasts and leaves-especially in C_3 plants (Sage *et al.*, 1989) - catalyzes the reaction between CO_2 and RuBP to produce triose phosphate. This product is then exported from the chloroplast stroma to the cytosol and converted into sucrose. Rubisco has a low catalytic rate per unit mass of protein, so high rates of CO_2 assimilation in C_3 leaves necessitate large quantities of Rubisco. Its concentration can reach up to 8 g/m^2, making up to 30% of the nitrogen in wheat leaves and up to 50% of the soluble protein. Actively metabolizing leaves thus have high nitrogen content. If nitrogen supply during leaf growth is insufficient to meet the demands for protein synthesis, the resulting amount of components will be inadequate for optimal CO_2 assimilation (Lawlor *et al.*, 1989b).

CO_2 assimilation relies not just on the quantity of components but on their activity, which is influenced by other factors. For instance, if Rubisco content is reduced, its activity may increase to compensate for the change. At lower temperatures, more Rubisco (along with other proteins and components) is needed to maintain a specific rate of CO_2 assimilation, as metabolic activity decreases due to slower enzyme reaction rates (Lawlor *et al.*, 1987a). Nitrogen deficiency significantly affects chloroplasts in terms of size, composition, and function. Chloroplasts in nitrogen-deficient plants are smaller and flatter compared to those in plants with sufficient nitrogen (Sow and Ranjan, 2021). They have fewer thylakoid membranes, which are poorly stacked, a smaller ratio of granal to intergranal thylakoids, and an increased proportion of stroma, as seen in rice (Laza *et al.*, 1993) and sugar beet (Kutik *et al.*, 1995).

Nitrogen Fixation: The Symbiotic Relationship

Nitrogen is essential for forming key biomolecules in plants, including proteins, nucleic acids, and chlorophyll. Despite making up 78% of the atmosphere, nitrogen in its gaseous form cannot be used directly by plants. To be utilized, it must be transformed into ammonium, nitrite, or nitrate ions. This conversion, known as nitrogen fixation (Fig. 4.2), can be assisted by plant growth-promoting rhizobacteria (PGPR).

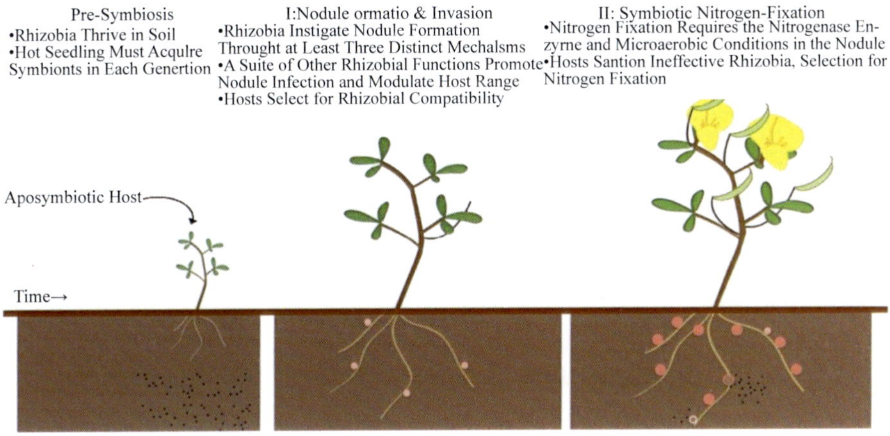

Fig. 4.2: Nitrogen fixation by root nodules of legume crop (Masson-Boivin and Sachs, 2018)

PGPR (Plant Growth-Promoting Rhizobacteria) can convert atmospheric nitrogen into ammonia through the enzyme nitrogenase (Gaby and Buckley, 2014). These bacteria can function either as symbionts or asymbionts. In symbiotic nitrogen fixation, nodules form on the roots of leguminous plants, where the bacteria reside (Sahoo *et al.*, 2024). Nitrogen fixation is a capability exclusive to prokaryotic organisms and is confined to a small but varied group of bacteria and blue-green algae, collectively known as diazotrophs, which belong to the eubacteria and archaebacteria kingdoms. In this process, diazotrophic symbionts (such as Rhizobium spp. and Anabaena azollae) live within specific structures provided by their plant hosts, such as root nodules in legumes or specialized cavities in the upper surface of the aquatic fern Azolla leaves (Giller, 2001).

Biological nitrogen fixation is a remarkable process. Dinitrogen, which is chemically triple-bonded, is introduced into the biosphere. Rhizobium bacteria invade the roots of compatible legume plants, leading to the formation of specialized root structures known as nodules. Inside these nodules, the bacteria use the nitrogenase enzyme complex to reduce N_2 to ammonia. This ammonia (NH_3) is quickly converted to ammonium (NH_4^+), which is then used to synthesize amino acids and proteins, temporarily stored in microbial or plant biomass (Pate, 1980). When these organisms die, their nitrogen-rich biomass contributes to the soil organic nitrogen (SON) pool. This nitrogen is later mineralized by soil microbes and released as forms such as nitrate (NO_3^-), which can be assimilated by other soil microbes and plant roots. In the European Union, biological nitrogen fixation in cultivated grain and forage legumes results in the fixation of approximately 800,000 tons of dinitrogen (N_2) each year, with major grain legume crops like soybean, pea, and faba bean accounting for about one-third of this total. In many tropical and subtropical agro-ecosystems, N_2 fixation by legumes plays a crucial role in the nitrogen economy of soils (Peoples and Herridge, 1990).

Nitrogen Assimilation and Metabolism

Legumes can obtain nitrogen from three main sources: NO_3^-, NH_4^+, and atmospheric N_2 through symbiotic fixation. Among these, NH_4^+ - N nutrition is the simplest metabolic process. Although NH_4^+ can be oxidized to NO3- in the soil, evidence suggests that in tropical soils, available NH_4^+ concentrations can be similar to NO_3^- levels (Uloro and Mengel, 1994). Nitrification may be hindered by high temperatures and low pH (Beck, 1983). Therefore, NH_4^+ nutrition can be particularly significant in tropical soils, while in temperate soils; nitrogen is mainly absorbed as NO_3^- (Mengel, 1991). The way nitrogen from these sources is incorporated into plant tissues involves different metabolic processes that significantly impact plant growth, especially when N_2 is fixed symbiotically by legumes.

a. Uptake and Assimilation of Ammonium and Nitrate

The uptake of NH_4^+ by root cells is driven by a plasmalemma-bound H^+ ATPase which creates both an electrical and pH gradient (Serrano, 1989; Sussman and Harper, 1989). As a cation, NH_4^+ is drawn into the cell passively through a specific transport system due to the negative charge (Bertl *et al.,* 1984). The partial depolarization of the plasma membrane and the release of protons during the incorporation of NH_3 into glutamate stimulate further proton release by the ATPase (Felle, 1989) (Fig 4). Consequently, legumes and other plants that rely on NH_4^+ nutrition will significantly acidify their rhizosphere (Schaller and Fischer, 1985).

In the cytosol, NH_4^+ is converted into glutamine by the enzyme glutamine synthetase (GS). Glutamine is then used by glutamine oxoglutarate aminotransferase (GOGAT, also known as glutamate synthase) in a transferase reaction to produce two molecules of glutamate by transferring the amino group from glutamine to oxoglutarate (Schubert, 1986). This process is the primary pathway for incorporating NH_4^+ into amino acids from all three nitrogen sources. The assimilation of NH_4^+ into oxoglutarate by glutamate dehydrogenase is less significant due to the high Km of this enzyme (Caba *et al.,* 1993). However, other studies (Sung and Chen, 1983) suggest that the activity of glutamate dehydrogenase may increase under water stress conditions. In contrast to NH_4^+, NO_3^- must be actively transported across the plasma membrane of root cells, working against an electrochemical gradient. This transport is believed to occur via a proton/nitrate symport system (Ullrich and Novacky, 1990). The symport of protons with NO_3^- results in proton consumption in the rhizosphere, thereby increasing the pH (Moorby *et al.,* 1985). This differs from the uptake of NH_4^+ or N_2 fixation, which typically leads to acidification of the rhizosphere (Bashan and Levanony, 1989).

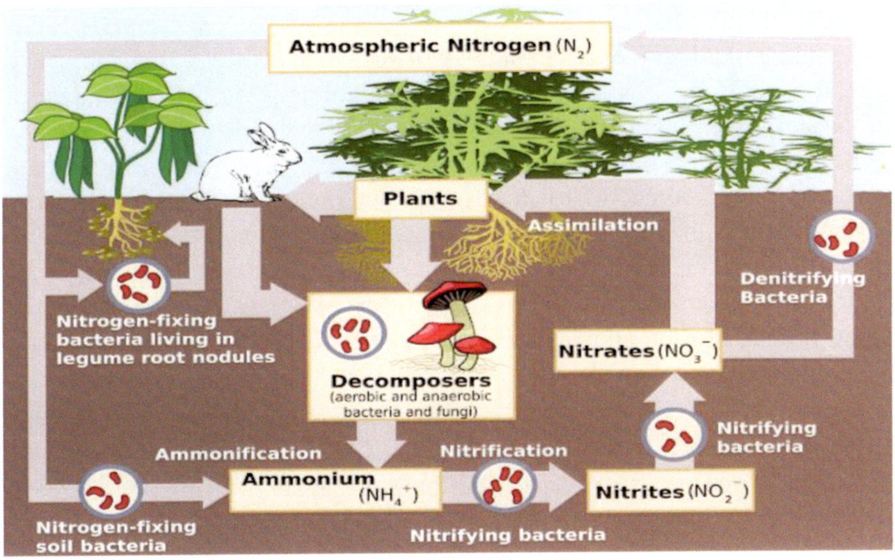

Fig. 4.3: Nitrogen cycle (Australian Earth Science Education)

Nitrate reduction (Fig. 4.3), driven by nitrate reductase, occurs in the cytosol and generates nitrite (NO_2^-). Since NO_2^- is further reduced to ammonium (NH_4^+) in the plastids, nitrate (NO_3^-) must pass through two membranes of the organelle envelope. This process involves the protonation of NO_2^- to form nitrous acid (HNO_2) which as a weak acid can easily diffuse through the membranes in its undissociated state (Heber and Heldt, 1981). The consumption of protons in the cytosol results in alkalinization, which in turn inhibits proton release by ATPase. This is one reason why NO_3^- nutrition is associated with an increase in rhizosphere pH (Moorby *et al.,* 1985; Allan and Raven, 1987).

b. Dinitrogen assimilation

The third source of nitrogen is atmospheric N_2, which legumes can access only through a symbiotic relationship with rhizobia. Within the Rhizobiaceae family, four genera are currently recognized (Elkan, 1992, Ranjan *et al.*, 2023) with Rhizobium and Bradyrhizobium being the most significant for agriculture. Effective strains of rhizobia induce the formation of nodules, specialized organs for N_2 fixation. These nodules originate from a secondary meristem in the root cortex tissue (Bauer, 1984; Werner *et al.*, 1992).

Nodules are composed of three primary tissues: the peripheral cortex, vascular bundles, and the central infection zone (Streeter, 1992, 1993). The vascular bundles, which are surrounded by the cortex and infected cells, connect to the root's phloem and xylem strands. Sucrose, which provides energy and carbon skeletons, is delivered into the infection zone via the phloem. There is some uncertainty regarding the exact pathway of sucrose transport into the infected cells. It is suggested that noninfected cells might assist in the movement of carbon and metabolites to the infected cells, given the frequent plasmodesmatal connections between infected and non-infected cells. The low concentration of sucrose and the absence of fructose in the nodule apoplast support the idea of symplastic unloading of sucrose from the phloem into the infected cells (Streeter, 1992). Evidence suggests that in root tissue, phloem unloading may occur via a symplastic pathway (Giaquinta *et al.,* 1983). Alternatively, sucrose might unload into the apoplast, where it could be hydrolyzed by invertase, allowing for selective uptake of hexoses by infected cells.

Hexoses are broken down through the glycolytic pathway, and organic acids are then transported into the bacteroids. Similar to mitochondria and chloroplasts, bacteroids are encased in a secondary membrane derived from the host cell's plasmalemma. This creates an organelle known as the 'symbiosome,' which resembles chloroplasts and mitochondria (Wemer *et al.,* 1992). The symbiosome facilitates N_2 fixation through the catalytic activity of nitrogenase. The nitrogenase-produced NH_4^+ is released into the cytosol via passive diffusion across the symbiosome membranes (Streeter, 1989).

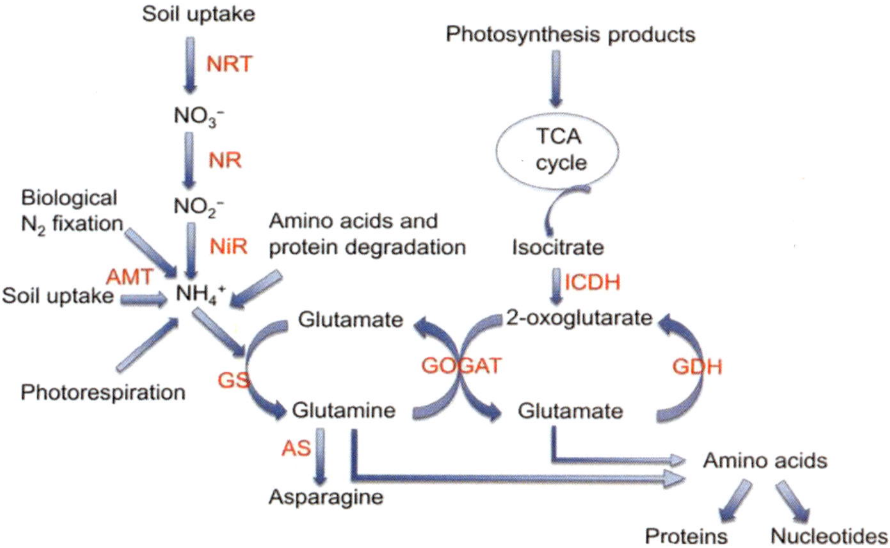

Fig. 4.4: Nitrogen assimilation (Lu *et al.*, 2016)

In the host cell cytosol, NH_4^+ is assimilated by the GS/GOGAT enzyme system (Mengel, 1994). Amides, or ureides in many tropical legumes, are then transported to the shoot through the xylem (Schubert, 1986). However, the specifics of how these compounds are exported from the infected cells to the xylem are not well understood. It is possible that active transport mechanisms play a role in maintaining low cytosolic amino acid concentrations in the infected cells, which is necessary for effective NH_4^+ assimilation. This, in turn, creates a diffusion gradient for NH_4^+ between the symbiosome and the cytosol. A steep gradient facilitates the release of NH_4^+ into the cytosol and helps prevent the inhibition of nitrogenase by NH_4^+ (Sung and Chen, 1983, Sow *et al.*, 2023).

Practical applications of C-N studies in grain legumes

Carbon and nitrogen assimilation are crucial processes in grain legumes, significantly impacting their growth, yield, and nutritional value.

Enhanced Crop Yields

1. **Carbon assimilation:** Grain legumes use carbon dioxide from the atmosphere, through photosynthesis, to produce carbohydrates which are vital for growth and seed development. The effective carbon assimilation can lead to increased biomass and better yields. Breeding and genetic modification for improved photosynthetic efficiency can therefore boost crop productivity.

2. **Nitrogen Assimilation:** Legumes form symbiotic relationships with rhizobium bacteria, which fix atmospheric nitrogen into a form usable by the plant. This process reduces the need for synthetic fertilizers, lowers production costs, and can lead to higher yields due to the improved nutrient availability.

Sustainable Agriculture

1. **Reduced Dependency on Chemical Fertilizers**: By improving nitrogen fixation through better legume management and breeding practices, farmers can reduce their reliance on chemical nitrogen fertilizers. This not only cuts costs but also minimizes environmental pollution from fertilizer runoff, which can contribute to issues like waterway eutrophication.
2. **Crop Rotation Benefits**: Incorporating legumes into crop rotation systems helps restore soil nitrogen levels, benefiting subsequent non-legume crops. This practice enhances soil health and reduces the need for synthetic nitrogen inputs.

Soil Health and Fertility

1. **Nitrogen Fixation:** Legumes enrich the soil with nitrogen, which is essential for soil fertility. This natural enrichment improves soil structure and nutrient content, supporting better growth for subsequent crops.
2. **Organic Matter Contribution:** The organic matter from legumes, including residues left after harvest, contributes to soil organic carbon. This helps in improving soil structure, water-holding capacity, and overall fertility.

Nutritional Quality

1. **Improved Protein Content:** Legumes are a significant source of plant-based protein. Efficient nitrogen assimilation boosts the protein content of the seeds, enhancing their nutritional value. This is beneficial for both human consumption and livestock feed.
2. **Enhanced Crop Quality:** High rate of carbon assimilation can lead to better growth and quality of the legumes, impacting not just yield but also the size, texture, and flavor of the grains.

Climate Change Mitigation

1. **Carbon Sequestration:** By enhancing carbon assimilation through improved photosynthesis, legumes can contribute to carbon sequestration in soils. This helps mitigate climate change by capturing atmospheric CO_2.

2. **Lower Greenhouse Gas Emissions:** Efficient nitrogen fixation reduces the need for synthetic fertilizers, which are associated with greenhouse gas emissions. This results in a lower carbon footprint for legume cultivation.

Economic Benefits

1. **Cost Savings:** Reduced need for nitrogen fertilizers and improved crop yields translate into lower production costs and higher profitability for farmers.
2. **Market Value:** High-quality, nutrient-rich legumes can command better market prices and meet the growing demand for plant-based proteins and sustainable agricultural practices.

Limitations

While carbon and nitrogen assimilation are vital for the growth and productivity of grain legumes, several limitations can impact their efficiency.

Environmental Factors

- **Light Intensity:** Insufficient light can limit photosynthesis, reducing carbon assimilation. Cloudy weather, shading from other plants or suboptimal light conditions can adversely affect legume growth and yield.
- **Temperature Extremes:** High temperatures can impair photosynthesis and nitrogen fixation, while low temperatures can slow down these processes, affecting overall plant health and productivity.
- **Water Stress:** Both drought and excessive water can hinder carbon assimilation by affecting photosynthetic efficiency and root health, which in turn impacts nitrogen fixation.

Soil Conditions

- **Nutrient Deficiency:** Even though legumes fix atmospheric nitrogen, the availability of other essential nutrients (such as phosphorus, potassium, and micronutrients) is crucial for optimal growth and nitrogen assimilation. Soil nutrient imbalances can limit legume performance.
- **Soil pH:** Extreme soil pH levels can negatively affect the activity of rhizobium bacteria, which are essential for nitrogen fixation. Acidic or alkaline soils may hinder the effectiveness of the symbiotic relationship.

Biological Constraints

- **Rhizobium Efficiency:** The effectiveness of nitrogen fixation depends on the compatibility and efficiency of the rhizobium strains present in the soil. Poor inoculation or ineffective rhizobium strains can limit nitrogen assimilation.

- **Plant Varieties:** Different legume varieties have varying capacities for carbon and nitrogen assimilation. Some may naturally perform better than others, and improving these traits through breeding or genetic modification can be challenging.

Pest and Disease Pressure

- **Pests:** Insects and other pests can damage legume plants, affecting their ability to photosynthesize and fix nitrogen effectively. Pest management strategies are essential to mitigate these impacts.
- **Diseases:** Fungal, bacterial, and viral diseases can interfere with carbon and nitrogen assimilation by damaging plant tissues or disrupting the root-rhizobium relationship.

Management Practices

- **Crop Management:** Poor agronomic practices, such as improper planting density, inadequate spacing, or incorrect fertilization, can limit the efficiency of carbon and nitrogen assimilation. Proper management is essential for optimizing these processes.
- **Competition with Weeds:** Weeds can compete with legumes for light, water, and nutrients, reducing their ability to assimilate carbon and nitrogen effectively.

Economic and Technological Constraints

- **Cost of Inputs:** Implementing advanced technologies or practices to enhance carbon and nitrogen assimilation may involve higher costs, which may not be feasible for all farmers.
- **Access to Knowledge and Resources:** Limited access to research, extension services, or modern agricultural practices can restrict the ability of farmers to optimize carbon and nitrogen assimilation in their legume crops.

Addressing these limitations requires a multifaceted approach, including improved management practices, better crop varieties, and advancements in agricultural technology.

Conclusion

Grain legumes, play a pivotal role in both human nutrition and sustainable agriculture. Their rich content of protein and essential nutrients makes them indispensable in diets worldwide, particularly for vegetarians. Their ability to fix atmospheric nitrogen through symbiotic relationships with rhizobia not only enriches soil fertility but also reduces the need for synthetic fertilizers, thus contributing to more environmentally friendly farming practices. Understanding the intricate processes of carbon and nitrogen assimilation in these legumes ranging from photosynthesis and nutrient uptake to nitrogen fixation reveals their complex biological functions and underscores their importance in achieving agricultural sustainability. By integrating legumes into crop rotations and employing them as both food and feed sources, we can enhance soil health, lower greenhouse gas emissions, and reduce reliance on non-renewable resources, aligning agricultural practices with principles of environmental stewardship and long-term productivity.

References

Allan, S., & Raven, J.A. (1987). Intracellular pH regulation in Ricinus communis grown with ammonium or nitrate as N source: The role of long distance transport. Journal of Experimental Botany, 38, 580-596.

Bashan, Y., & Levanony, H. (1989). Effect of root environment on proton efflux in wheat roots. Plant and Soil, 119, 191-197.

Bauer, W.D. (1981). Infection of legumes by rhizobia. Annual Review of Plant Physiology, 32, 407-449.

Beck, T. (1983). Die N-Mineralisierung von Böden im Laborbrutversuch. Journal of Plant Nutrition and Soil Science, 146, 243-252.

Bertl, A., Felle, H., & Bentrup, F. W. (1984). Amine transport in Riccia fluitans. Cytoplasmic and vacuolar pH recorded by a pH-sensitive microelectrode. Plant Physiology. 76, 75-78.

Caba, J. M., Llucb, C., & Ligero, F. (1993). Genotypic differences in nitrogen assimilation in Vicia faba: Effect of nitrate. Plant and Soil, 151, 167-174.

Elkan, G. H. (1992). Taxonomy of the rhizobia. Canadian Journal of Microbiology, 38, 446-450.

Felle, H. (1989). pH as a second messenger in plants. In: Second Messengers in Plant Growth and Development, Alan Liss Inc. New York, pp 145-166.

Foyer, C., Ferrario-Mery, S., & Noctor, G. (2001). Interactions between carbon and nitrogen metabolism. In: Lea PJ, Morot-Gaudry J-F, eds. Plant nitrogen. Berlin: Springer-Verlag. pp 237-254.

Gaby, J. C., & Buckley, D. H. (2014). A comprehensive evaluation of PCR primers to amplify the nifH gene of nitrogenase. Plos One, 9(3), e93883.

Giaquinta, R. T., Lin, W., Sadler, N. L., & Franceschi, V. R. (1983). Pathway of phloem unloading of sucrose in corn roots. Plant Physiology, 72, 362-367.

Giller, K.E. (2001). Nitrogen Fixation in Tropical Cropping Systems, second ed. CABI Publishing, Wallingford.

Heber, U., & Heldt, H. W. (1981). The chloroplast envelope: Structure, function, and role in leaf metabolism. Annual Review of Plant Physiology, 32, 139-168.

Kutik, J., Natr, L.H.H., Derks, D., & Lawlor, D. W. (1995). Chloroplast ultrastructure of sugarbeet (Beta vulgaris L.) cultivated in normal and elevated CO_2 concentrations with two contrasted nitrogen supplies. Journal of Experimental Botany, 46(12), 1797-1802.

Lawlor, D. W., Boyle, F. A., Young, A. T., Keys, A. J., & Kendall, A. C. (1987a). Nitrate Nutrition and Temperature Effects on Wheat: Photosynthesis and Photorespiration of Leaves. Journal of Experimental Botany, 38(188), 393-408.

Lawlor, D. W., Kontturi, M., & Young, A. T. (1989b). Photosynthesis by flag leaves of wheat in relation to protein, ribulose bisphosphate carboxylase activity and nitrogen supply. Journal of Experimental Botany, 40(210), 43-52.

Lawlor. D. W. (2002). Carbon and nitrogen assimilation in relation to yield: mechanisms are the key to understanding production systems. Journal of Experimental Botany, 53(370), 773-787.

Laza, R.C., Bergman, B., & Vergara, B. S. (1993). Cultivar differences in growth and chloroplast ultrastructure in rice as affected by nitrogen. Journal of Experimental Botany, 44, 1643-1648.

Lu, J., Zhang, L., Lewis, R. S., Bovet, L., Goepfert, S., Jack, A. M., et al. (2016). Expression of a constitutively active nitrate reductase variant in tobacco reduces tobacco-specific nitrosamine accumulation in cured leaves and cigarette smoke. Plant Biotechnology Journal, 14(7), 1500-1510.

Masson-Boivin, C., & Sachs, J. L. (2018). Symbiotic nitrogen fixation by rhizobia—the roots of a success story. Current Opinion in Plant Biology, 44, 7-15.

Mengel, K. (1991). Available nitrogen in soils and its determination by the "Nmln-method" and by electro ultrafiltration (EUF). Fertilizer Research, 28, 251-262.

Mengel, K. (1994). Symbiotic dinitrogen fixation-its dependence on plant nutrition and its ecophysiological impact. Zeitschrift für Pflanzenernährung und Bodenkunde, 157(3), 233-241.

Mondal, D., Sadhukhan, T., Latif, I. A., & Datta, S. N. (2015). Quantum chemical investigation of thermochemistry in Calvin cycle. Journal of Chemical Sciences, 127, 2231-2240.

Moorby, H., Nye, P. H., & White, R. E. (1985). The influence of nitrate nutrition on H^+ efflux by young rape plants. Plant and Soil. 84, 403 - 415.

Pate, J. S. (1980). Transport and partitioning of nitrogenous solutes. Annual Review of Plant Physiology and Plant Molecular Biology, 31, 313-340.

Peoples, M. B., & Herridge, D. F. (1990). Nitrogen fixation by legumes in tropical and subtropical agriculture. Advances in Agronomy, 44, 155-223.

Ranjan, S., Kumar, S., Dutta, S. K., Sow, S., Kumar, S., & Sushant. (2023). Long-term organic amendment application improves soil fertility status, nutrient accumulation and crop productivity under rice-wheat cropping system. Communications in Soil Science and Plant Analysis, 54(18), 2579-2589.

Ranjan, S., Sow, S., Kumar, S., Ghosh, M., Roy, D. K., Dutta, S. K., & Pramanick, B. (2022). Enhancing Nutrient Use Efficiency and Productivity of Cereals through Site Specific Nutrient Management. Journal of Cereal Research, 14(3), 229-242.

Rochester, I. J., Peoples, M. B., Constable, G. A., & Gault, R. R. (1998). Faba beans and other legumes add nitrogen to irrigated cotton cropping systems. Australian Journal of Experimental Agriculture. 38(3), 253-260.

Sage, R. F., Sharkey, T. D., & Seemann, J. R. (1989). Acclimation of Photosynthesis to elevated CO_2 in five C_3 species. Plant Physiology, 89, 590-596.

Sahoo, R., Sow, S., Ranjan, S., Dharminder, Kumar, R., Roy, D. K., et al. (2024). Unveiling the potential of plant growth promoting rhizobacteria (PGPR) in phytoremediation of heavy metal. Discover Applied Sciences, 6(6), 324.

Schaller, G., & Fischer, W. R. (1985). pH changes in the rhizosphere of maize and peanut roots. Journal of Plant Nutrition and Soil Science, 148(3), 306-320.

Schubert, K. R. (1986). Products of biological nitrogen fixation in higher plants: Synthesis, transport, and metabolism. Annual Review of Pfant Physiology, 37, 539-574.

Serrano, R. (1989). Structure and function of plasma membrane ATPase. Annual Review of Plant Physiology and Plant Molecular Biology, 40, 61-94.

Singh, R. J., Chung, G. H., & Nelson, R. L. (2007). Landmark research in legumes. Genome, 50, 525–537.

Sow, S, & Ranjan, S. (2021). Precision Nutrient Management: As a tool to enhance Nutrient Use Efficiency. Agriblossom, 1(7), 3-9.

Sow, S., Singh, G., Ghosh, M., Dutta, S. K., Mandal, N., Kumar, S., & Ranjan, S. (2023). Nitrogen-management strategy through leaf-colour chart and SPAD meter for optimizing the productivity in irrigated wheat (Triticum aestivum). Indian Journal of Agronomy, 68(2), 219-222.

Sow, S., Ranjan, S., Padhan, S.R., Nath, D., & Kumar, N. (2024). Agroforestry and Soil Carbon Sequestration: A Nexus for System Sustainability. In: Kumar, S., Alam, B., Taria, S., Singh, P., Yadav, A., Arunachalam, A. (eds) Agroforestry Solutions for Climate Change and Environmental Restoration. Springer, Singapore, pp 103-126.

Stagnari, F., Maggio, A., Galieni, A., & Pisante, M. (2017). Multiple benefits of legumes for agricultural sustainability: an overview. Chemical and Biological Technologies in Agriculture, 4, 2.

Streeter, J. G. (1992). Analysis of apoplastic solutes in the cortex of soybean nodules. Physiologia Plantarum, 84, 584-592.

Streeter, J. G. (1993). Translocation – A key factor limiting the efficiency of nitrogen fixation in legume nodules. Physiologia Plantarum, 87, 616-623.

Streeter. J. G. (1989). Estimation of ammonium concentration in the cytosol of soybean nodules. Plant Physiology, 90, 779-782.

Sung, J. M., & Chen, C. L. (1983). The effect of water stress on enzymes of ammonia assimilation in cytosol of soybean nodules. Journal of the Agricultural Association of China, 121, 20-27

Sussman, M. R., & Harper, J. F. (1989). Molecular biology of the plasma membrane of higher plants. Plant Cell, 1, 953-960.

Ullrich, C. I., & Novacky, A. J. (1990). Extra and intracellular pH and membrane potential changes induced by K^+, Cl^-, H_2PO_4, and NO_3 uptake and fusicoccin in root hairs of Limnobium stoloniferum. Plant Physiology, 94, 1561-1567.

Uloro, Y., & Mengel, K. (1994). Response of ensete (Ensete ventricosum W.) to mineral fertilizer applications under representative growth conditions in the vicinity of Hossaina, Southwest Ethiopia. Fertilizer Research, 37, 107-113.

Wemer, D., Ahlborn, B., Bassarab, S., Kape, F., Kinnback, A., Mellor, R. B., M6rschel, E., Muller, P., Parnike, M., Schmidt, P., & Schultes, A. (1992). Nodule development and nitrogen fixation in the Rhizobium/Bradyrhizobium system. In: Mengel K and Pilbeam DJ (eds) Nitrogen Metabolism of Plants, Proceedings of the Phytochemical Society of Europe 33. Oxford Science Publications, UK, pp 17-30.

Zahran H. H. (1999). Rhizobium-legume symbiosis and nitrogen fixation under severe conditions and in an arid climate. Microbiology and Molecular Biology Reviews, 63(4), 968-89.

5

Climate Resilience, Mitigation and Adaptation Strategies of Grain Legumes

Lakshman[1*], Jayshree[1], Suwa Lal Yadav[2] and Anu Nawhal[3]

[1]*Department of Agronomy, Chandra Shekhar Azad University of Agriculture & Technology, Kanpur, Uttar Pradesh, India*

[2]*Assistant Professor, Raffles University, Neemrana, Alwar, Rajasthan, India*

[3]*Department of Agronomy, Naini Agricultural Institute, Sam Higginbottom University of Agricultural, Technology and Sciences, Prayagraj, Uttar Pradesh, India*

**Corresponding author's email: lakshman7518@gmail.com*

Abstract

Global food production needs to double by 50% by the middle of the 21st century to ensure food and nutritional security for 9 billion people. Climate change, particularly atmospheric CO_2 and temperature, has a complex impact on sustainable food security. Climate change is a significant constraint on legume production globally, which is a crucial food, feed, and fodder source. Legumes fix atmospheric nitrogen through bacteria in root nodules, reducing greenhouse gas emissions and boosting soil C-sequestration. Climate change positively impacts grain legumes by allowing them to allocate above-ground photosynthates into below-ground parts (nodules), improving rhizospheres' activities, plant biomass and nutrient use efficacy. However, temperature can also negatively affect grain legumes, leading to flower abortion and abnormal pod filling. However, climate models predict more extreme events in the future, affecting legume growth and development. Heat stress, elevated CO_2 concentration, drought, and rainfall variability impact legume crop production, requiring adaptation options. Results show that legumes with C_3 fixation pathways show higher photosynthesis, reduced photorespiration, biomass production, and water use efficiency under atmospheric CO_2. However, temperature increases lead to faster development, shorter life cycles, and lower yields. Legumes can cope with atmospheric CO_2 up to 1000 ppm by storing sucrose and starch. Protein synthesis and water stress also help legumes adapt to changing climates. To achieve targeted yields in changing climatic conditions, agronomic and genetic interventions must be *integrated. Targeted breeding programs should select high-yielding, biotic stress-resistant, and abiotic stress-tolerant genotypes adapted to atmospheric CO_2 and temperature conditions. Integrating grain legumes with desirable traits can provide resilience against climate change.*

Keywords: *Climate change, Grain legumes, Nutritional security, Resilience, Rhizosphere, Water Use Effciency*

Introduction

Climate change is a significant concern for agricultural development, as humanity faces a challenge in producing enough food for an additional three billion people within the next 50 years. This is due to adverse conditions such as shrinking arable lands, becoming more calorie-intensive food habits, scarcity of water and nutrients, and changing climate, particularly in areas where extra food production is needed, such as the semi-arid tropics. Prediction models predict a 2-4°C increase in temperature over the next century, with atmospheric CO_2 concentration increasing at a rate of 1 ppm/year over the last 50 years. Precipitation is predicted to increase by 10-20% depending on region, and climatic events are expected to increase, leading to more frequent heat stress.

Climate change will primarily affect regions where cropping conditions are already under environmental pressure, particularly in terms of water. Resilience of crop yield under harsh conditions, such as the semi-arid tropics, is a problem that must be addressed today. The excessive use of natural resources, including fossil fuel burning, deforestation, intensive farming without crop rotation, industrial waste, and landfills, has led to a significant increase in greenhouse gases (GHGs) such as carbon dioxide, methane, and nitrous oxide. Over the last five decades, the world has experienced an average rise in CO_2 concentration by 155%, with the highest concentration in 2015 and a spike in global temperature by 0.85°C. Over the last five years, global average temperature and atmospheric CO_2 concentration have increased by 0.2°C and 20%, respectively (Sow *et al.*, 2023).

Drought is the most widespread and yield-limiting abiotic stress in many crops, particularly in legumes. To adapt crops to climate change, it is crucial to improve the tolerance of crops to water deficits. Climate change is predicted to exacerbate regional and global water scarcity, with two-thirds of the world population already experiencing severe water scarcity. Agriculture is highly sensitive to environmental fluctuations resulting from global climate change, impacting flowering phenology, water availability, soil fertility, pathogen spread, and host susceptibility. These vulnerabilities pose a serious threat to global food and nutritional security. There is an urgent need to identify crops that can mitigate climate change threats, with grain legumes emerging as a suitable group with promising future prospects. Grain legumes are a valuable resource for mankind due to their high protein content, essential elements, dietary fibre, vitamins, carbohydrates, sugars, vitamins, essential mineral elements, mono and polyunsaturated fatty acids, dietary fibres, folic acids, and non-nutritional bioactive components. Incorporating pulses in daily diets can prevent numerous health problems like cardiovascular diseases, type 2 diabetes, and some forms of cancer, and help reduce the risk of obesity.

Grain legumes, which cover 81 million hectares and produce over 92 million tonnes globally, are cultivated in countries like China, Myanmar, Canada, Australia, Brazil, Argentina, USA, and Russia. India is the largest consumer of these legumes, accounting for nearly 1/4th of global annual production. These legumes are adapted to various environmental conditions, including cool to warm seasons. India is home to various grain legumes, including chickpea, pigeonpea, lentil, field pea, cowpea, horsegram, greengram, blackgram, chickling pea, and kidney bean. Grain legumes play an important role in providing ecosystem services, such as biological nitrogen fixation (BNF) capacity, deep root systems, low input requirements, propensity to survive in problem soils, and withstand abiotic stresses. They are popular choices for farming communities in intensive cereal-based cropping systems, which are one of the inputs intensive practices responsible for greenhouse gas emissions, ground water pollution, and other second-generation problems.

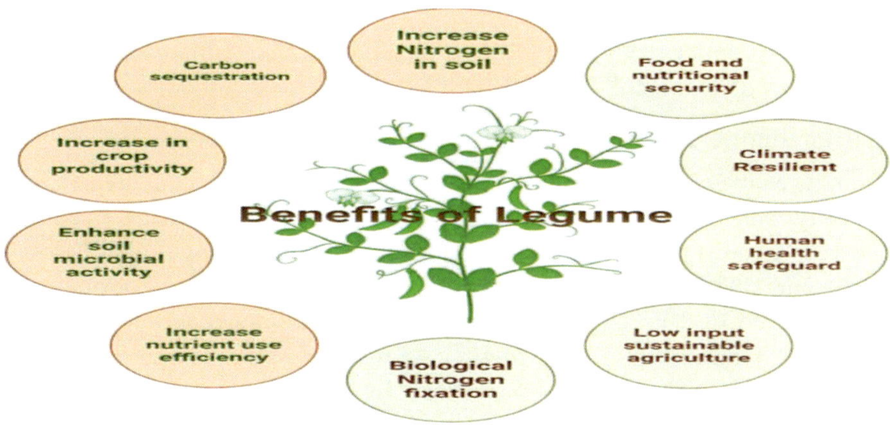

Fig. 5.1: Schematic representation of the benefits of legumes for humans and soil (Dave *et al.*, 2024)

Legumes have an important role in carbon sequestration due to their leaf fall, deep root system, symbiotic nitrogen fixation, and carbon rich root exudates. They significantly improve soil carbon retention in reduced tillage conditions due to higher crop biomass and moderating the rate of carbon mineralization compared to cereals. Legumes translocate a large part of their photosynthetic carbon as root exudates and various lignin-rich compounds, contributing largely to carbon sequestration and reducing carbon footprint. Legumes enrich the soil with nitrogen for subsequent crops and curtail global CO2 emission up to 50% that emits from N-fertilizer (urea) industries. Nitrous oxide emission is also found to be lower from legumes (1.02 kg N_2 O–N ha^{-1} $year^{-1}$) as compared to N-fertilized cereals (2.71 kg N_2 O–N ha^{-1} $year^{-1}$).

Plants with C_3 pathways are favoured by elevated carbon dioxide due to adaptive strategies like higher rate of photosynthesis, sharp increase in soluble sugars and starch, rise of growth-promoting hormones like indole acetic acid and symbiotic nitrogen fixation (SNF), making grain legumes superior to others. The most important mechanisms by which legume cope up with atmospheric CO_2 is symbiotic nitrogen fixation, which serves as a source of carbon and energy for the nitrogenase activity.

Climate change impacts water use efficiency (WUE) and legume productivity, particularly in arid and semi-arid ecosystems. Moisture stress, combined with atmospheric T and atmospheric CO_2, has a significant impact on legume growth and productivity. Climate change has also altered the geographical distribution of pests, leading to increased secondary pest infestations and loss of valuable produce. Legumes can adapt to climate change positively or negatively, depending on internal physiological adjustments and crop management practices. Understanding the adaptation strategies and phenological triggers for legumes under changing climate is challenging due to their indeterminate nature (Sahoo *et al.*, 2024).

Rising Temperature Threat to Grain Legumes

Climate change is significantly impacting the global food system, human food security, and human health. Average temperatures have already risen 1°C above pre-industrial levels and are expected to increase by 1.5°C between 2030 and 2052. Under optimistic projections, global legume crop yields are expected to decrease by about 10 %. However, under business-as-usual scenarios, temperatures may rise 2.5 - 4.8°C by 2100, causing median yield losses between 20 and 49% for major legumes like beans, chickpeas, cowpeas, lentils, and soybeans.

Nearly all legume species show marked declines in key physiological processes beyond 30-35ºC, with extreme heat exposure over 40capable of inducing complete reproductive failure. Intensive agricultural and industrial development are changing the agricultural production and productivity of legume crops, leading to a decreasing productivity. Climate change presents a serious threat to legumes, as they can provide better yields. The greenhouse gas effect is increasing on crops, especially legume crops, making it difficult to achieve yield targets. Temperature and CO_2 levels exert direct and indirect effects on the physiological and biochemical characteristics of grain legumes. Threshold temperature ranges of some leguminous crops for heat stress are given in Table 1.

Table 5.1: The heat stress threshold temperature ranges of some leguminous crops

Legume Crop	Threshold Temperature (°C)
Chickpea	15–30
Pigeon pea	18–30
Cowpea	18–28
Soybean	23–26
Groundnut	30–35
Urd bean	25–35
Mungbean	28–35
Pea	15–25
Lentil	15–30

Under Threat: Global Goals towards Future Security

The global population is expected to reach 9.7 billion by 2050, leading to a 51% rise in the demand for plant-based proteins. Food systems are crucial for sustainable development, particularly zero hunger and poverty by 2030. Climate change is a major challenge for legume crops, which already provide 33 % of the total dietary protein nitrogen for humans. However, legume yields in top producer regions like sub-Saharan Africa, South Asia, and Australia are likely to fall below baseline levels to meet future nutritional demands.

A multitude of species are on the verge of extinction as a result of human activity, with future population expansion and economic development providing extraordinary extinction risks to many species, notably large mammals in tropical Africa, Asia, and South America. However, these threats are not inevitable. Proactive international efforts to increase crop yields, minimize land clearing and habitat fragmentation, and protect natural lands could increase food security in developing nations and preserve much of Earth's remaining biodiversity.

Building Resilience: Climate Adaptation in Legumes

Food security has evolved to include nutritional security with nutritional demand. Vegetable legumes, such as peas, soybeans, fava beans, broad beans, and cowpeas, contribute to long-term crop production and food security, boosting both rural and urban economies. Leguminous vegetables, such as peas, soybeans, fava beans, broad beans, and cowpeas, are rich in micronutrients and can prevent diseases like cardiovascular and metabolic diseases. They also have beneficial properties, including minerals and vitamins, and play a role in metabolic reactions and coenzymes.

However, climate change affects vegetable production, and without urgent efforts to improve legume climate resilience, progress on malnutrition reduction, protein transition, smallholder farmer livelihoods, and sustainable agriculture worldwide may be at risk. The literature highlights the risks faced by legume producers globally and emphasizes the need for priority adaptations and mechanisms to manage climate uncertainty, enabling legumes to sustainably fulfil current and future food security roles.

A. Unbalanced Advantages: Downside of CO_2

The Industrial Revolution has led to a rise in atmospheric CO_2 concentration, which directly impacts crop plants by decreasing their yield and affecting their response to future increases in CO_2. Understanding how plants respond and adapt to future increases is crucial for understanding their current responses and potential adaptations. Over the past 20 years, extensive research has focused on the effects of CO_2 increases on plant physiology and development, providing valuable insights into plant adaptations.

i) **Boosting Legume yield with elevating CO_2:** As atmospheric CO_2 concentrations rise above 400 ppm, the "CO_2 fertilization effect" directly stimulates photosynthesis and water retention in C_3 plants like legumes. Elevated CO_2 enhances legume photosynthetic rates by 55% and water use efficiency by 48% across diverse environments and genetic material. However, long-term free-air CO_2 enrichment studies show that the actual yields increase by just 18% under elevated levels compared to ambient CO_2 levels. The accelerated photosynthesis from CO_2 fertilization often fails to translate to higher yields due to other limiting factors affecting carbon allocation and reproductive development. Despite the potential metabolism aspects affected by CO_2, there is clear evidence only for effects on the Rubisco and stomatal movement in the range of $[CO_2]$ relevant to global change (i.e., 270–1000 μmol mol^{-1}). Plants can only perceive changes in atmospheric concentration through tissues exposed to the open air, and only the photosynthetic organs have direct contact with the atmosphere. The direct increase in photosynthesis due to CO_2 elevation results from two properties of Rubisco terrestrial C_3 plants.

Elevated CO_2 Impact on Legumes

- CO_2 increases carboxylation velocity due to its close proximity to atmospheric concentration.
- CO_2 inhibits oxygenation reaction, leading to glycolate production and photorespiration.
- This effect enhances net carbon CO_2 uptake efficiency by reducing photorespiratory CO_2 loss and diverting ATP and NADPH from photorespiratory metabolism to photosynthetic assimilation.
- The stomatal aperture's response to CO_2 variations remains a mystery.
- Long-term elevated CO_2 affects legumes as shown in Table 2.

Table 5.2: Examples of effects of long-term atmospheric CO_2 on seed yield and biomass production in a selection of grain legumes

Legumes	Effect
Pigeon pea (*Cajanus cajan*)	Shows 29% increase in total dry weight and yield across years
Mung bean (*Vigna radiata*)	Increase up to 14% in seed yield and 11% in total biomass
Peanut (*Arachis hypogaea*)	25% increase across year in seed yield and 16% increase in dry weight across years; no change in root dry weight
Lentil (*Lens culinaris*)	Increase by 18–13% in seed yield 50% increase in ABG biomass across cultivars between flowering and maturity 18% increase in ABG biomass across cultivars at flowering
Soybean (*Glycine max*)	Up to 20% increase in seed yield across the year depending on the cultivars Increase by 13–32% in ABG biomass depending on cultivars across year
Field pea (*Pisum sativum*)	Seed yield increases 26% across cultivars and 19–33% increase in dry weight biomass across cultivars

ii) **CO_2 Fertilisation in Changing Climate:** Rising CO_2 levels have been found to have a stimulatory effect on legume crop yields, but this effect is expected to plateau around 550 ppm across bean, soybean, and peanut varieties. However, atmospheric CO_2 is projected to rise beyond 700 ppm by 2050 and exceed 900 ppm by 2100, causing temperatures to surpass optimal legume growing thresholds. The threshold of 550 ppm in CO_2 concentration refers to the level at which significant responses in plant yield are observed. The increased heat-induced flower abortion and restricted productive duration will overshadow any enhancements in operational photosynthetic capacity. The CO_2 fertilization effect may slightly boost legume yields over the coming decade, but it cannot protect global legume production from other existential threats posed by rising emissions, such as extreme temperature, water availability, and soil and nutrient availability. Climate models project intensified intra-seasonal precipitation variability under climate change, posing challenges for legume farmers in planting timing, fertilizer application, and harvest scheduling. Excess rainfall increases the risk of fungal disease outbreaks, as shown in the case of foliar Bean Common Mosaic Virus, with infection rates doubling under simulated future high precipitation regimes.

B. Frequent Droughts threats to Legume

Legumes are highly sensitive to drought, with severe drought significantly affecting their final yield. Drought affects germination percentage, photosynthetic activity, flowering time, and reproductive stages, leading to poor seed set and decreased sink activity. Drought is one of the uncontrolled and unpredictable phenomena that directly affects legume crops due to global climate change. It disturbs plant morphology, physiology, and growing period. Advanced approaches for dry environments, such as drought tolerance traits, innovative breeding, and water-efficient practices like drip irrigation and mulching, are crucial for enhancing legume productivity in harsh environments. Studies on drought effects, mechanisms, and management strategies may help manage the devastating effects of drought stress and develop drought-tolerant genotypes in dry environments. Climate projections suggest that total annual precipitation will decline across major legume-producing areas by the middle of this century, leading to decreased legume yields. Some legume crop stages are affected by drought, as listed in Table 3.

Table 5.3: Some legume crop stages that are affected by drought

Legume	Growth Stage	Yield Loss (%)
Soybean	Reproductive phase	46–71
	Grain Filling stage	42
	Pod Set	45–50
Chickpea	Anthesis Stage	27–40
	Reproductive Phase	45–69
	Ripening Stage	49–54
Common Bean	Flowering Stage	49
	Reproductive Phase	58–87
	Pod filling Phase	40
Lentil	Reproductive phase	24
	Pod development	70
Pigeon Pea	Reproductive Phase	40–55
Faba Bean	Grain Filling stage	68
Cowpea	Reproductive Phase	60
	Pod filling Phase	29

C. Legumes face heat threats

Legumes possess various morphological attributes, such as deep root systems, higher water use efficiency, nitrogen fixing ability, and low water and nutrient requirements. They often experience heat resistance during the reproductive stage, leading to the inhibition of photosynthesis and reduction in pollen fertility. Improving water relation characteristics could be one of the best strategies to address heat tolerance.

Rising temperatures and declining moisture availability pose a significant risk to legume crop losses, with peanut plant mortality tripled following simultaneous exposure to drought and 40°C heat. Flower abortion in heat-stressed soybean plants also soars from 27% to 74% alongside soil drying. Male reproductive development in higher plants is sensitive to heat stress at all growth stages, with high-temperature stress resulting in lower seed set due to male sterility in most legume crops. Temperatures more than 30°C reduce stigmatic receptivity, stigmatic pollen germination, stigma and style growth, and ovule penetration. Breeding legume cultivars with enhanced drought or heat tolerance shows strong potential to counteract shifting precipitation regimes. However, the adaptation pace must accelerate rapidly as up to 70 % of global legumes face precipitation declines.

Emerging genomic tools can enable better assessment of available genetic diversity for climate resilience within legume germplasm collections. Investing in irrigation infrastructure is critical, as supplemental water during episodic moisture deficits may reduce legume flower abortion by 29-39 % even under moderated heat stress. Empowering farmers to access improved irrigation will be essential to stabilize legume yields and farm incomes amidst intensifying precipitation variability.

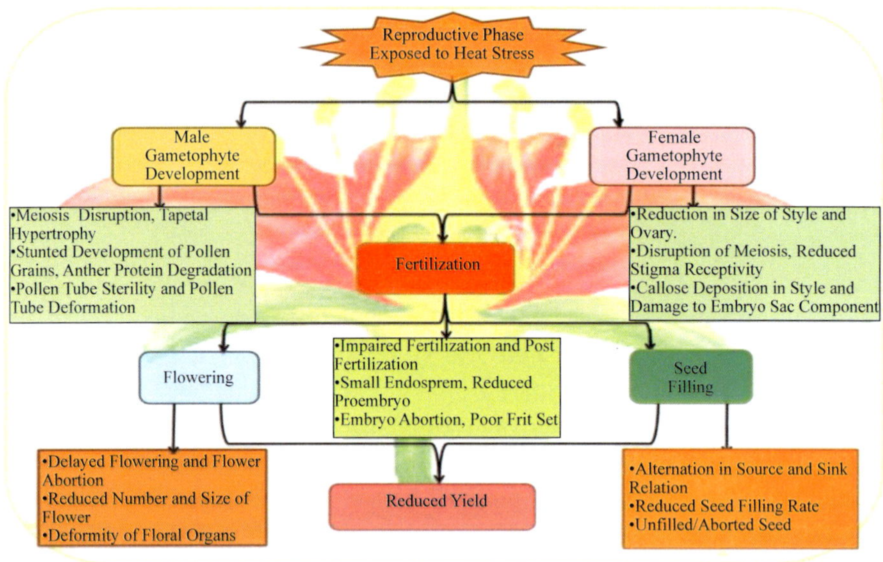

Fig. 5.2: Effects of heat stress during the reproductive phase (at different functional stages) (Sita *et al.*, 2017)

Heat stress during the flowering stage poses a significant threat to the quality and productivity of legumes. The temperature for legumes during flowering varies according to the season, with winter temperatures ranging from 12-18°C and summer temperatures between 21-28°C. Grain legumes are most sensitive to heat stress during the flowering, pod set, and seed filling growth phases. Short periods of extreme temperatures of 35°C or above increase flower or pod abortion by 22-45 % across legume species, with chickpeas showing near 100% loss of reproductive capacity beyond 36°C. Heat stress also affects grain development by disintegrating the tapetum layer, reducing nutrient supply to the microspore, leading to premature anther dehiscence, which controls the synthesis and distribution of carbohydrates to grain, ultimately reducing yield. Transient heat waves also reduce grain-fill duration, affecting legume yields. Episodes above 34°C during the seed development stage decreased harvested dry bean seed weights up to 61% relative to ambient conditions. Soybeans also experienced a 5% decline in seed oil and protein contents over the entire growing season.

Table 5.4: Effects of heat stress on food legumes at different stages of growth

Food Legumes	**Threshold Temperature(°C)**	**Heat Stress (Day/Night)**	**Effects**
			Reduced oxygen evolution and photochemical
Pea	15–25	30/25	energy storage; shutdown of PSI reaction center;
			interrupted electron donation by PS II's OEC
			(Oxygen-Evolving Center).
Lentil	15–30	38/23	Decreased electron flow during photosynthesis.
			Damaged structure and function of associated
			enzymes and proteins;
Chickpea	15–30	35/16	reduced stigma responsiveness; inhibited
			RuBisCO and sucrose metabolism in leaves;
			disturbed PSII.
Peanut	30–35	38/22	Impaired photosystem II; decreased pollen
			production
Black gram	25–35	35	Reduced photosynthesis
			Reduced chlorophyll content and photosynthesis;
Soybean	26	38/30	pollen tube growth and yield; shrunken pollen;
			damaged PSII.
			Reduced sugar in leaves due to decreased sucrose
Mungbean	28–35	>40/25	synthesis enzymes and RuBisCO activity;
			efficiency of photosynthesis impaired
Cow pea	18–28	36/27	Tapetal cell deterioration and anther indehiscence

D. Uncertainty due to Pest and Pathogen

Climate change is a significant factor affecting the growth and development of insect pests, which are cold-blooded insects with body temperatures similar to the environment. These pests can alter their physiology and behaviour due to changes in host plant resistance and interactions with other insects and natural enemies. Increasing temperatures accelerate the reproductive cycles of poikilothermic crop pests, allowing additional generations per season and amplifying their population sizes. Vegetable legumes, a rich source of protein, are affected by various plant insect pests and pathogens, causing yield losses directly in legume crops. Warmer conditions also expand the viable spatial ranges for legume pests, allowing influxes into historically unaffected areas.

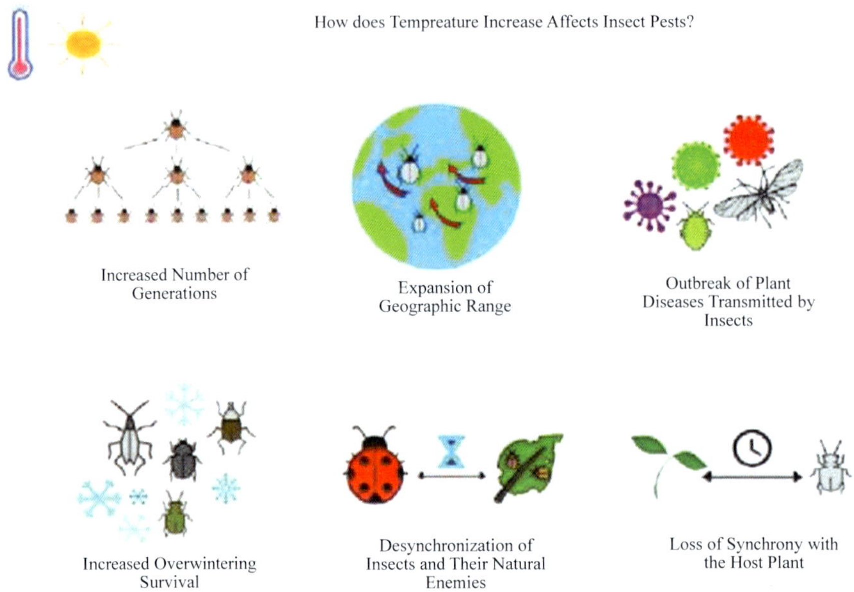

Fig. 5.3: Impacts of increasing temperature on agricultural insect pests (Kumar *et al.*, 2024)

Multiple aggressive legume pathogens, such as *Sclerotinia* stem rot, charcoal rot, and *Cercospora* leaf spot, are exhibiting accelerated growth and heightened virulence under elevated temperatures and precipitation variability. Intermittent rainfall patterns can have a significant impact on insect species that hibernate in soil, leading to prolonged water stagnation and floods. Insect survival is threatened, and intense rain and flooding can wash away insect eggs and larvae.

The proliferation of existing legume pests and pathogens, along with the possible emergence of novel threats under changing climate, presents a significant challenge for food security. Integrating climatic risk factors into pest management frameworks and improving landscape diversification through cropping system rotations and targeted agroforestry interventions may help buffer pest pressures amplified under climate shifts. Breeding durable multi-trait insect and pathogen resistance should be a top priority.

E. Soil Fertility Decline: A Silent Crisis

The symbiotic relationship between Rhizobia bacteria and atmospheric nitrogen fixation in legume crops is affected by drought stress, which impairs nitrogen fixation through nodule damage and carbohydrate partitioning changes. Soil drying may decrease symbiotic nitrogen inputs, affecting sustainable intensification goals without external fertilizer amendments. The loss of functional nodules and fixed nitrogen during early crop establishment under recurrent heat waves can lead to compounded soil nutrient deficiency throughout the cropping cycles. Climate extremes may erase any transient boosts in operational photosynthetic productivity from CO_2 fertilization. The intersection of rising temperatures, shifting rainfall, and current extremes can unravel legume crop productivity through exacerbated soil nutrient deficits. The combined drought and heat events during the seed filling phase decreased nitrogen fixation rates in faba bean, chickpea, and lentil plots by 29-76% relative to ambient conditions. Prolonged elevated temperatures also hasten native soil organic matter decomposition and release substantial CO_2 stores back into the atmosphere, further impeding farm system resilience.

F. Impact of Climate Change on Nodulation

Climate change has been extensively studied in relation to legume physiology, particularly nodule growth and performance. Nodule formation and nitrogen fixation process involve multiple steps, but the impact of atmospheric CO_2 on individual steps is difficult to assess. In general, atmospheric CO_2 has a positive response on nodule growth and physiology, which has been documented for multiple legumes. Climate change majorly impacts nodulation or nodule growth by altering nodule physical parameters (nodule biomass and nodule number per plant) and changing nodule biological characteristics (nitrogenase activity, leghaemoglobin content, and N_2 fixation). However, the impact of atmospheric CO_2 on nodulation is highly linked with crop species, bacterial symbiont, plant nutritional status, soil physicochemical conditions of soil, and environmental conditions.

Exposure time of atmospheric CO_2 has a significant impact on nodule growth and development. Short-term exposure promotes nodule functions like nitrogenase activity, N2 fixation, and whole-plant N status, while prolonged exposure modifies nodule development, such as nodule biomass and nodule number per plant.

Nodule biomass is also positively influenced by P-supplementation. Under atmospheric CO_2 (550 ppm) with addition of P, nodule biomass significantly increased by 46 and 49% in chickpea and field pea, respectively. However, results are not always significant and the extents of change in different legumes are hard to ascertain. Climate smart adaptive strategies by legumes include higher photosynthetic rate and allocation of carbon in the roots due to atmospheric CO_2, which eventually boosts nitrogenase (Nase) activity and SNF. Biochemical parameters like leghaemoglobin content and Nase activity were influenced positively with the rise of gaseous concentration. High temperature stress is likely to impact gene regulations related to nodulation and rhizobium infection. The activity of rhizobia is highly temperature-sensitive and optimum at 20-25°C. Minor fluctuations in soil temperature even by less than 2°C can ravage the symbiotic relationship by hampering root growth and rate of infection.

Under atmospheric temperature conditions, accumulation of toxic nitrogenous compounds and inappropriate carbohydrate supply to the root system hamper normal root hair growth, degenerates nod gene activity, and eventually makes the delicate symbiotic relationship between crop-microorganism fragile. Experiments showed that temperature higher than 39°C alters the gene functioning related to flavonoid production in beans and soybean, impeding the normal infection process and restricting the growth of root hairs considerably.

In inference, developing climate resilient agro-technologies for suitable agro-ecosystems will help legumes adapt and fulfil the grain demand for mankind.

Climate Change Mitigation via Legumes and Their Microbiome

Climate change is accelerating or enhancing the incidence of various abiotic and biotic stresses on agricultural and forest productivity. By changing different aspects of agriculture management, we can lower the risk of climate change over plant-derived productivity, including food, fibre, and bioenergy. Legumes (Fabaceae) are the third largest family of flowering plants, with over 20000 species, playing central roles in human and animal food and plant-based protein production. They emit fewer greenhouse gases, allow more carbon sequestration in soils, and save fossil energy inputs by reducing nitrogen fertilizer needs. In intercropping systems, legumes may facilitate circulation of soil nutrients, water retention and availability, and break pest and disease cycles, depending on the climate, soil, and management of agro-ecosystems in which they are included.

Mutualistic relationships of legumes with bacteria or fungi have been investigated intensively, usually as bi-partite interactions. Their collective contribution to plant growth and performance appears to be affected by crop genotype, microbial interactions, and environmental variables. The interaction with the microbiome, as well as the outcome of microbe-microbe interactions, enable plants to receive a wide range of benefits like increased tolerance against biotic and abiotic stresses, biological control of pathogens, and enhanced nutrient acquisition and growth. The legume holobiome (legume and its microbiome as a unit) is a promising tool to mitigate the climate change impact while addressing the need for healthy and nutritious food for a rapidly growing population. The challenge is to understand the dynamics of the legume-intercrop-microbiome-soil system and use those insights as a solid basis for developing new integrated technologies and practices.

Mitigating the impacts of climate change on grain legumes is crucial for sustaining agricultural productivity and food security. Grain legumes, such as soybeans, chickpeas, lentils, and beans, are highly sensitive to climatic factors. Implementing strategies to mitigate these impacts involves a combination of agricultural practices, breeding programs, and policy initiatives. Here are some effective mitigation strategies:

1. Breeding and Genetic Improvement

- **Heat and Drought Tolerance:** Developing varieties that can withstand higher temperatures and drought conditions is crucial. Breeding programs focus on traits like deep root systems, efficient water use, and heat-resistant flowering and pod formation.
- **Pest and Disease Resistance:** Climate change can lead to the emergence of new pests and diseases. Breeding for resistance can help maintain productivity under these new threats.
- **Nutrient Use Efficiency:** Varieties that can efficiently use available nutrients can reduce the need for fertilizers, minimizing environmental impacts.

2. Agronomical Practices

- **Crop Rotation and Intercropping:** Rotating legumes with cereals or intercropping with other crops can improve soil health, reduce pest and disease incidence, and enhance resilience to climate extremes.
- **Conservation Agriculture:** Practices like minimal tillage, maintaining soil cover, and crop diversification can improve soil structure and moisture retention, making the system more resilient to climate variability.
- **Integrated Pest Management (IPM):** Using biological controls, resistant varieties, and cultural practices to manage pests can reduce reliance on chemical pesticides, which are often less effective under changing climatic conditions.

3. Water Management

- **Efficient Irrigation Systems:** Implementing drip or sprinkler irrigation systems can optimize water use efficiency, ensuring that crops receive adequate water even during periods of scarcity.
- **Rainwater Harvesting and Storage:** Capturing and storing rainwater can provide a supplementary water source during dry spells.
- **Soil Moisture Conservation:** Mulching, cover cropping, and other practices that reduce soil evaporation can help conserve moisture.

4. Climate-Smart Soil Management

- **Organic Matter Addition:** Adding organic matter, such as compost or manure, can improve soil structure and water-holding capacity.
- **Cover Cropping:** Planting cover crops can protect the soil from erosion, improve soil health, and provide additional biomass for soil organic matter.
- **pH Management:** Maintaining optimal soil pH can enhance nutrient availability and uptake, which is crucial for crop resilience.

5. Technological Innovations

- **Precision Agriculture:** Utilizing sensors, drones, and satellite imagery to monitor crop health, soil conditions, and water needs can optimize inputs and reduce waste.
- **Decision Support Systems:** Implementing tools that provide real-time advice on planting dates, irrigation schedules, and pest management can improve farm-level decision-making.
- **Genomic Tools:** Employing genomic selection and marker-assisted breeding can accelerate the development of climate-resilient varieties.

6. Policy and Institutional Support

- **Incentives for Sustainable Practices:** Governments can provide subsidies or incentives for farmers adopting sustainable and climate-resilient practices.
- **Research and Development:** Continued investment in research on climate-resilient crops and farming systems is essential.
- **Extension Services:** Strengthening extension services to disseminate knowledge and technologies to farmers is critical for widespread adoption.

7. Community-Based Adaptation

- **Participatory Breeding Programs:** Involving farmers in the selection and breeding process can ensure the development of varieties that meet local needs and conditions.
- **Knowledge Sharing:** Facilitating the exchange of knowledge and best practices among farmers can enhance collective resilience.
- **Local Climate Adaptation Plans:** Developing community-level plans that address specific local climate challenges can provide targeted support and resources.

8. Economic Diversification

- **Value Addition and Processing:** Developing value-added products from grain legumes can provide additional income streams for farmers.
- **Market Access and Supply Chains:** Strengthening supply chains and improving market access can help stabilize incomes and reduce vulnerability to climate shocks.

By integrating these strategies, farmers can enhance the resilience of grain legume production systems to the adverse effects of climate change, ensuring food security and sustainability.

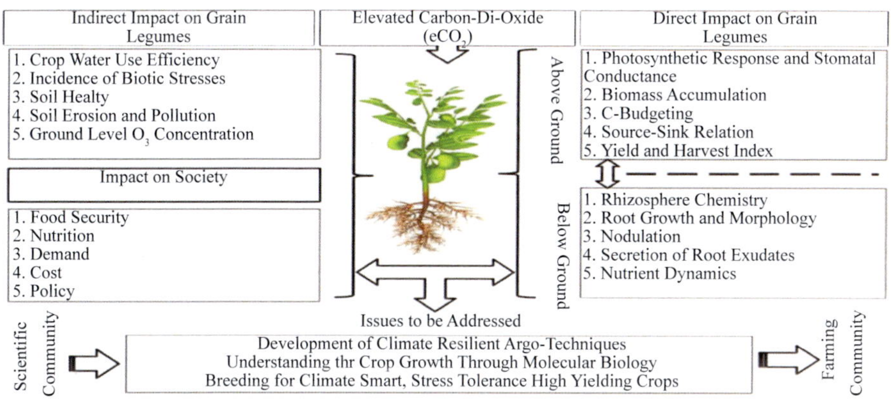

Fig. 5.4: Direct and indirect impact of grain legumes and future recommendations (Dutta *et al.*, 2022)

Cutting Edge Technologies to Hasten Legume Climate Resilience

A. **Effective Rhizobia Strain Selection and Fertilisation for Adaptation:** To maintain nitrogen availability in the face of climate change, it is crucial to develop legume cultivars that can sustain effective nodulation under intermittent moisture stress and heat waves. Rapid genomic screening of diverse germplasms can enable this. Inoculation with elite

Rhizobium strains exhibiting superior heat tolerance is an accessible tactic for producers in the future. Incorporating judicious inorganic fertilizer applications may also be necessary to stabilize yield goals, ensuring symbiotic fixation is not inhibited.

B. **Unleashing Potential: Accelerating Climate Resilient Legume Development:** Whole-genome sequencing, precision phenotyping platforms, and machine learning algorithms can accelerate the delivery of climate-adapted legume varieties to producers' fields. High-throughput genomic selection allows simultaneous screening of complex traits like heat tolerance across extensive germplasm collections. These techniques, combined with expanding public accessibility to international crop gene banks via digital interfaces, democratize breeding material access. However, additional financing is needed to integrate identified hardy lineages into pre-breeding pipelines and commercial variety advancement systems targeting climate-vulnerable geographies in developing countries.

C. **Gene Editing Offers Potential for Climate-Protective Legume Innovations:** Gene editing techniques like CRISPR-Cas can develop transgenic, climate-resilient legume varieties at accelerated timelines. However, these applications are primarily focused on model organisms like soybean. To democratize climate-resilience benefits for vulnerable producers and consumers, further tool development and regulatory environments should be directed towards nutrition-critical legumes like cowpeas, faba beans, and lentils. Strategic gene editing may offer rapid, precise, and revolutionary adaptation solutions, particularly for complex, quantitative climate tolerance traits not easily manipulated through traditional crossing.

D. **Option for Expanding Climate Risk Protection:** The use of plant-microbe interactions is gaining attention for improving agricultural resilience. Custom microbiome bio stimulant products and biofertilizers with elite rhizobacteria can enhance legume germination, nutrient uptake, and pathogen resistance under climate stress. These bio-solutions are accessible and can be used by legume producers. Microbial seed coatings encapsulate nitrogen-fixing rhizobia, while foliar biopesticides use antifungal bacteria or essential oils to manage yield losses from virulent diseases. Expanding commercialization and farmer education on these microbiomes and biopesticide innovations is crucial for addressing climate risk management in legume production.

E. **Heat-Tolerant Properties for Climate Change:** Climate change adaptation is a top priority for developing legume cultivars tolerant to heat and aridity extremes. However, the complex polygenic architecture of heat stress response presents challenges for rapid variety improvement. Integrating quantitative trait loci mapping, genome-wide association studies, and genomic selection approaches can expedite the delivery of hardy legume germplasms. Plant breeding alone is not enough to elicit adaptation, as heat-tolerant lines will struggle to deliver productivity gains under mid-century temperatures.

Conclusion

Climate change poses significant threats to sustainable legume-supported food systems worldwide. However, emerging innovations, such as omics-empowered crop improvement platforms, gene editing techniques, and plant microbiome harnessing, can unlock legumes' unique environmental services for productive, nutrition-secure, and ecologically multifunctional agriculture on a climate-disrupted planet. Grain legumes are the best alternatives for anchoring protein transitions across societies worldwide. Realizing climate resilience is an urgent multidimensional challenge, with rising temperatures, shifting precipitation extremes, and increased climate variability destabilizing agroecosystems. To address these challenges, adaptive interventions such as heat/drought-tolerant legume cultivar development, enhanced climate-informed agronomic management, refined biological nitrogen fixation solutions, strengthened climate pest/disease monitoring systems, and supportive agricultural development policies targeting smallholder legume producer resilience must be implemented. Adhering to ambitious emission reduction targets is also crucial, as global temperature trajectories beyond 1.5°C increase risk exposure for vulnerable farming communities. The climate resilience of leguminous vegetables now plays an indispensable role in determining the sustainability of the entire global food system.

References

Dave, K., Kumar, A., Dave, N., Jain, M., Dhanda, P.S., Yadav, A., & Kaushik, P. (2024). Climate Change Impacts on Legume Physiology and Ecosystem Dynamics: A Multifaceted Perspective. Sustainability, 16, 6026.

Dhaliwal, S.K., Talukdar, A., Gautam, A., Sharma, P., Sharma, V., & Kaushik, P. (2020). Developments and prospects in imperative underexploited vegetable legumes breeding: A review. Int. J. Mol. Sci., 21, 9615.

Dutta, A., Trivedi, A., Nath, C.P., Gupta, D.S., & Hazra, K.K. (2022). A comprehensive review on grain legumes as climate-smart crops: Challenges and prospects. Environmental Challenges, 7, 100479.

FAO. (2022). Crops and Climate Change Impact Briefs: Climate-Smart Agriculture for More Sustainable, Resilient, and Equitable Food Systems; Food & Agriculture Organization: Rome, Italy.

Khatun, M., Sarkar, S., Era, F.M., Islam, A.M., Anwar, M.P., Fahad, S., Datta, R., & Islam, A.A. (2021). Drought stress in grain legumes: Effects, tolerance mechanisms and management. Agronomy, 11, 2374.

Kumar, J., Choudhary, A.K., Gupta, D.S., & Kumar, S. (2019). Towards exploitation of adaptive traits for climate-resilient smart pulses. Int. J. Mol. Sci., 20, 2971.

Kumar, N., Hazra, K.K., Nath, C.P., Praharaj, C.S., & Singh, U. (2018). Grain legumes for resource conservation and agricultural sustainability in South Asia. In: Legumes For Soil Health and Sustainable Management. Springer, Singapore, pp. 77–107.

Kumar, S., Kaur, R., Kaur, N., Bhandhari, K., Kaushal, N., Gupta, K., Bains, T., & Nayyar, H. (2011). Heat-stress induced inhibition in growth and chlorosis in mungbean (Phaseolus aureus Roxb.) is partly mitigated by ascorbic acid application and is related to reduction in oxidative stress. Acta Physiol. Plant, 33, 2091–2101.

Kumar, S., Kumar, P., Suryawanshi, D., & Omar, V. (2024). Effect of Global Warming on the Insect-Pest of Agricultural Crops in book Advances in Agricultural Entomology. Pp. 69-83.

Prasad, P.V.V., Craufurd, P.Q., Kakani, V.G., Wheeler, T.R., & Boote, K.J. (2001). Influence of high temperature during pre-and post-anthesis stages of floral development on fruit-set and pollen germination in peanut. Funct. Plant Biol., 28, 233–240.

Sahoo, R., Sow, S., Ranjan, S., Dharminder, Kumar, R., Roy, D. K., et al. (2024). Unveiling the potential of plant growth promoting rhizobacteria (PGPR) in phytoremediation of heavy metal. Discover Applied Sciences, 6(6), 324.

Sharma, M., Kaushal, R., Kaushik, P., & Ramakrishna, S. (2021). Carbon farming: Prospects and challenges. Sustainability, 13, 11122.

Shrestha, S. (2019). Effects of climate change in agricultural insect pest. Acta Sci. Agric., 3, 74–80.

Singer, S.D., Chatterton, S., Soolanayakanahally, R.Y., Subedi, U., Chen, G., & Acharya, S.N. (2020). Potential effects of a high CO_2 future on leguminous species. Plant-Environ. Interact., 1, 67–94.

Sita. K., Sehgal, A., Rao, H.B., Nair, R.M., VaraPrasad, P.V., Kumar, S., Gaur, P.M., Farooq, M., Siddique, K.H.M., Varshney, R.K., & Nayyar, H. (2017). Food Legumes and Rising Temperatures: Effects, Adaptive Functional Mechanisms Specific to Reproductive Growth Stage and Strategies to Improve Heat Tolerance. Frontiers in Plant Science, 8, 1658.

Sow, S., Ranjan, S., Behera, B., Ghosh, M., Kumar, S., & Dutta, S. K. (2023). Maintaining agricultural sustainability through carbon footprint management. Current Science, 125(9), 939-944.

Vadez, V., Berger, J., Warkentin, Asseng, Ratnakumar, Rao, K., Gaur, P., Nathalie, G., Munier-Jolain, Larmure, M., & Voisin, A. (2012). Adaptation of grain legumes to climate change: a review. Agronomy for Sustainable Development, 32(1), 31-44.

Vijayalakshmi, D., Priya, J.R., Vinitha, A., & Ramya, G. (2024). Interactive effects of elevated CO_2 with combined heat and drought stresses on the physiology and yield of C_3 and C_4 plants. J. Crop Sci. Biotechnol., 27, 1–16.

6

Efficient Water Use and Productivity of Grain Legume Cropping Systems: Recent Advances and Technological Interventions

Sweeti Kumari[1]*, Mukesh Kumar Sinha[2], Rina Roy[3] and Shweta Kumari[4]

[1]*Department of Agronomy, Veer Kunwar Singh College of Agriculture, Dumraon Buxar, Bihar, India*

[2]*Department of Agricultural Economics, Veer Kunwar Singh College of Agriculture Dumraon, Buxar, Bihar, India*

[3]*Department of Agronomy, Dr. Kalam Agricultural College, Kishanganj, Bihar, India*

[4]*Department of Soil Science and Agricultural Chemistry, Dr. Kalam Agricultural College, Kishanganj, Bihar, India*

**Corresponding author's email: sneha54545454@gmail.com*

Abstract

Yield of grain legumes are constrained by available water. Thus, it is crucial to understand traits influencing water uptake and the efficiency of using water to produce biomass. Global comparisons at specific locations reveal that water use of different grain legumes is very similar, which indicates that water use efficiency varies over a wide range due to differences in biomass and yield. Moreover, yield increases more per millimetre of water used in cool season grain legumes than warm season species. Although greater contrasts have been observed across species and genotypes, agronomic factors need to be taken into account when scaling those studies to field-level responses. Conservative water use strategies in grain legumes such as low stomatal conductance as approximated by low photosynthetic carbon isotope discrimination reduces yield potential, whereas temporal adjustments of stomatal conductance within the growing season and in response to environmental factors helps to optimize the trade-off between carbon gain and water loss. Furthermore, improved photosynthetic capacity, reduced mesophyll conductance, reduced boundary layer, and re-fixation of respired CO_2 were identified as traits that are beneficial without water deficit, but also *under terminal and transient drought. Genotypic variability in some grain legume species has been observed for several traits that influence water use, water use efficiency and yield, including root length and the temporal pattern of water use, but even more variation is expected from wild relatives. Conservative traits are needed under conditions of terminal drought to help maintain soil moisture until the pod-filling period, but profligate traits, if tightly regulated, are important under conditions of transient drought in order to profit from short intermittent periods of available soil moisture.*

Keywords: *Grain legumes, Terminal drought, Transient drought, Water use efficiency*

Introduction

Agriculture remains highly sensitive to the environmental fluctuations resulting from the global climate change (Lenka and Lal, 2012). Climate change impacts several aspects of agricultural production systems such include alteration in flowering phenology, water availability, soil fertility and erosion, increase in pathogen spread and host susceptibility (Rosenzweig and Hillel, 1995). These climate change induced vulnerabilities, possess a serious threat to the global food and nutritional security. There is an urgent need to identify crops that can mitigate the threats of climate change. Grain legumes can emerge as one such suitable group of crops having prospects for future. Grain legumes are nature's precious gift to mankind as these are rich in protein (16–50%).essential elements, dietary fibre (10–23%) and vitamins (Maphosa and Jideani, 2017). Apart from protein, grain legumes are store house of various nutritional components such as: carbohydrates, sugars, vitamins, more than 15 essential mineral elements and mono and polyunsaturated fatty acids (Wang *et al.*, 2009). Growing grain legumes in crop rotations is a sustainable way to meet future demands for food and protein (Foyer *et al.*, 2016). Their ability to introduce atmospheric nitrogen into agricultural systems through symbiotic N_2 fixation, reducing the need for nitrogen fertilizers to improve soil health and structure, to break disease cycle and the high protein content of grains used for both humans and animal feed are well known (Hauggaard-Nielsen *et al.*, 2008; Bampidis and Christodoulou, 2011). However, total production and land area planted to grain legumes is only a fraction of that compared with cereals (Foyer *et al.*, 2016). Low and variable yields (Cernay *et al.*, 2015) constrain their adoption in farming systems (Millan *et al.*, 2006; Anderson *et al.*, 2016), especially due to increased risk when combined with high input costs, such as fungicides and herbicides. Improvements in yields have been slow or declining for most grain legumes (Foyer *et al.*, 2016), indicating that greater investment and research is needed to close the yield gap and improve yield stability. Higher grain yields are associated with greater water supply for transpiration (T), better capture and use of water for carbon assimilation, transpiration efficiency (TE), and converting more of the carbon assimilated (biomass) into grain, harvest index (HI), where yield is a function of T×TE×HI (Passioura, 1977). This book chapter concludes that conservative traits are needed under conditions of terminal drought to help maintain soil moisture until the pod-filling period, but profligate traits, if tightly regulated, are important under conditions of transient drought in order to profit from short intermittent periods of available soil moisture.

Important Grain Legume Crops of India

As rainfed crops, legumes are mostly mixed or intercropped with cereals or long-season and widely-spaced crops. Cropping systems research has identified - compatible crops for intercropping and the optimum proportions and spatial arrangements that yield maximum intercropping advantage. Chickpea or gram (*Cicer arietinum* L.) and pigeonpea or "tur" (*Cajanus cajan*) which together account for 61% of the total pulse production, are the principal grain legumes; chickpeas are grown in the post-rainy season (Oct.-March) and pigeonpea are planted in the rainy season (June- Oct.). Other legumes grown exclusively in the post-rainy season are lentil (*Lens culinaris* Medic), "khesari" or grasspea (*Lathyrus sativus* L.) and peas (*Pisum sativum* L.). Mungbean or green gram (*Vigna radiata*) "urd" or black gram (*Vigna mungo*) and cowpea (*Vigna sinensis*) are grown in both seasons but the post-rainy season crop is possible only in the warmer parts of the country. Where irrigation is available these can also be grown in the summer. Other grain legumes grown in the rainy season with limited regional importance are field beans (*Lablab purpureus* L.) Sweet moth bean (*Vigna acantifolia*), cluster beans (*Cyamopsis tetragonolobus*) and soybean (*Glycine max*). Groundnut (*Arachis hypogaea*) is the major oil seed legume grown in both seasons but in the post-rainy season the crop is mostly confined to the irrigated areas of peninsular India.

Growing Conditions and Indeterminate Growth of Grain Legumes

Grain legumes can be divided into two groups depending on the timing of their productive period and temperature requirements for growth; they experience transient or terminal drought depending on the climate of their growing region and availability of water (Table 1). When grown on stored soil moisture with only limited additional rainfall during the growing season, insufficient starting soil moisture to last the season or water use (WU), that exhausts the stored soil moisture can lead to terminal drought. Following the differentiation into terminal and transient drought of Berger *et al.* (2016), warm season and cool season grain legumes grown in Mediterranean-type climates are more likely to be exposed to transient drought, while cool season grain legumes grown on stored soil moisture in subtropical climates are more likely to face terminal drought. Moreover, on a global basis, yield reduction of grain legumes due to drought, strongly depends on the soil texture (Daryanto *et al.,* 2017), thus increasing the importance of soil type on soil water availability. The growth habit of most grain legumes is indeterminate, with vegetative growth continuing while flowering, which can result in long developmental stages, variable maturing times, and a long critical period determining yield (Lake and Sadras, 2014). While these factors can be problematical for intensive farming, indeterminate growth permits response to rainfall or other favourable conditions at any time. In addition, the extended period of flowering (and pod set) compensates for flower abortion due to short-term transient stresses or grazing, thus giving indeterminate plants phenotypic plasticity. On the other hand, flowering and early pod-filling periods are most vulnerable to stresses, so indeterminate crops face an increased likelihood that their critical developmental phases will experience variable stress condition.

Climate of Dominant Growing Zones

Mediterranean type climates (Koppen–Trewartha type Csa and Csb), characterized by mild, but wet winters and dry summers (Kottek *et al.*, 2006). Humid sub-tropical climates (Koppen–Trewartha type Cfa and Cwa), characterized by hot summers with summer dominant rainfall and cool drier winters (Belda *et al.*, 2014). Semi-arid to low humid tropics or temperate regions, in Central and South America, Africa, and Europe (Sinha, 1977) covering a range of climates such as Koppen–Trewartha type Aw, Cfa, Dfa, and Dfb Main growing season. Planted in autumn and harvested in late spring. Planted in the post-rainy season, summer crops mostly grown on stored soil moisture.

Table 6.1: Common grain legume growing condition (Lawn, 1989)

Particulars	**Cool season**		**Warm season**
Climate of growing region	Mediterranean type climates (Koppen– Trewartha type Csa andCsb), characterized by mild, but wet winters and dry summers (Kottek *et al.*, 2006)	Humid subtropical climates (Koppen–Trewartha type Cfa and Cwa), characterized by hot summers with summerdominant rainfall and cool drier winters (Belda *et al.*, 2014)	Semi-arid to low humid tropics or in temperate regions in Central and South America, Africa, and Europe (Sinha, 1977) covering a range of climates such as Köppen– Trewartha type Aw, Cfa, Dfa, and Dfb
Main growing season	Planted in autumn and harvested in late spring	Planted in the post-rainy season	Summer crops
Water availability	In-season intermittent rainfall	Mostly grown on stored soil moisture	In-season rainfall or irrigation
Drought type	Mostly transient drought, but also terminal drought	Terminal drought	Short-term transient drought
Plant developmental stage when drought occurs	Vegetative, reproductive, and pod filling	Reproductive and pod filling	Any

Critical Growth Stage for Grain Legumes

Drought of different intensity is experienced in rainfed areas at various growth stages of legumes. Response to limited irrigation has been observed in most of the grain legumes. Among various crops, common bean (*Phaseolus vulgaris* L.; French bean) was found to be more responsive to irrigation followed by pea. The success of mung bean as a catch crop (during summer months) in the rice-wheat system is solely dependent upon adequate supply of irrigation plants (Keller *et al.,* 2000). Late sown chickpea in sequence with rice also needs irrigation compared to the normal sown crop, probably due to restricted root growth with late sowing. Pre flowering or flower initiation and post podding has been found to be the most critical stage in most of the legumes. However, the initial soil profile moisture and soil types largely determine the requirement of subsequent irrigation. Similarly, excess moisture or water logging reduces oxygen concentration in the rhizosphere and thus affects biological nitrogen Fixtation activity and nutrient availability with consequent yield reduction. Therefore, it is imperative to provide good drainage, especially in low-lying areas

Comparing WU of Grain Legumes and Cereals

Despite higher carbon requirements for protein synthesis and N_2 fixation, not all grain legumes use more water (or have lower yields) than cereals, which has been attributed to legumes having shallower roots and higher harvest index (Angadi *et al.*, 2008). Similarly, on a global scale, legumes cannot be classified as using more water than cereals. Cool season grain legumes, in particular, exhibit a similar or lower range of WU compared with wheat. While yield is generally positively related to WU (except see the case for faba, cluster, and moth bean which are not significant), the yield responses to water are far lower for legumes (3.3–8.2 kg ha^{-1} mm) than those for cereals (Table 2) and the proportion of yields that can be explained by amount of water used varies over a wide range (0–83%) (Table 2). The spread in data indicates that factors other than WU also drive yields (French and Schultz, 1984).

Table 6.2: Relationship between water use and yield for a range of cool (A) and warm (B) season grain legume species

	Species	R^2	Significance	Slope (kg ha^{-1} mm water used)
Cereal	Bread wheat (*Triticum aestivum*)	0.69	****	11.4 ± 1.2 a
	Durum wheat (*T. durum*)	0.80	****	12.8 ± 1.7 a
Cool season	Chickpea (*Cicer arietinum*)	0.56	****	8.2 ± 0.9 a
	Lentil (*Lens culinaris*)	0.40	****	3.9 ± 0.6 b
	Field pea (*Pisum sativum*)	0.22	****	5.6 ± 1.5 c
	Faba bean (*Vicia faba*)	0.22	ns	ns
Warm season	Soybean (*Glycine max*)	0.59	***	3.1 ± 0.7 a
	Peanut (*Arachis hypogaea*)	0.54	*	5.9 ± 1.9 a
	Pigeon pea (*Cajanus cajan*)	0.57	**	2.5 ± 0.7 a
	Mung bean (*Vigna radiata*)	0.83	****	4.1 ± 0.4 a
	Cowpea (*Vigna unguiculata*)	0.76	***	3.7 ± 0.6 a
	Cluster bean (*Cyamopsis tetragonoloba*)	0.39	ns	ns
	Moth bean (*Vigna aconitifolia*)	0.02	ns	ns

The data are from a global data set compiled by Cernay *et al.* (2016) from field experiments published in 173 articles between 1967 and 2016. Only species with >10 data points were considered for the analysis. Yield and water use are positively correlated for most species, with the correlations tending to be stronger for warm season species. Slopes were compared within a species group, and different letters indicate a significant difference (P=0.196) of 3.3 kg ha^{-1} mm water used (pooled slope). Hence, warm season legumes use more water to produce the same yields, which is expected due to higher soil evaporation at higher temperatures. Data for wheat are shown for comparison.

Water Use Efficiency in Grain Legumes

Water use efficiency (Yield/Evapotranspiration) is the outcome of an entire suite of plant and environmental processes operating over the life of a crop to determine both Yield (Y) and Evapotranspiration (ET). Consequently, biomass production per unit Evapotranspiration (ET), has been used extensively as an interim measure of water use efficiency. Evapotranspiration (ET) comprises non- productive evaporation (E) of water from the soil surface and productive transpiration (T) of soil stored water by the plant. Evaporation of free water from leaf surfaces adds to non-productive evaporation (interception evaporation) (Ali, 2008). Crop grown in a sunny and hot climate needs more water per day than the same crop grown in a cloudy and cooler climate. There are, however, apart from sunshine and temperature, other climatic factors which influence the crop water need. These factors are humidity and wind speed. When it is dry, the crop water needs are higher than when it is humid. In windy climates, the crops will use more water than in calm climates. The highest crop water needs are thus found in areas which are hot, dry, windy and sunny. The lowest values are found when it is cool, humid and cloudy with little or no wind. From the above, it is clear that the crop grown in different climatic zones will have different water needs and thus water use efficiency also varies accordingly (Ali *et al.,* 2012). It was noticed by several researchers that dicot plants are more efficient than monocots, similarly C_3 plan are performed better than C_4 plants (Keller *et al.*, 2000).

WUE at Field Level

From an agronomic viewpoint on a field level, WUE is defined as:

Biomass per water used (WUE_b) = B/WU (1)

Grain yield per water used (WUE_g) = G/WU (2)

where B is (aboveground) biomass, G is grain yield, and water use (WU) is the total over the whole growing season, which integrates water losses via evapotranspiration (ET), namely crop transpiration (T) and soil evaporation (E). WU at this level is usually calculated from soil moisture content at the start of the season less soil moisture at the end of the season plus rainfall and/ or irrigation. A more mechanistic field-level WUE (Sinclair *et al.*, 1984) is:

$$WUE_g = (1\text{-}E/ET) \times HI \times k/e_i - e_a \quad (3)$$

Where, harvest index (HI)= (G/B), atmospheric vapour pressure (e_a) and leaf intercellular vapour pressure (e_i). As e_i is generally assumed to be at saturation, the term $e_i – e_a$ corresponds to ambient vapour pressure deficit (VPD), with $e_i – e_a$ being the mean VPD for times with transpiration. Based on some simplifying assumptions, the tr$_i$anspiration coefficient k is a constant defined as:

$$k = 1.6 \times m \times h \times (1 - p_i/ p_a) \times p_a \times LD/ LT \quad (4)$$

Where, LD and LT are leaf area indices separating the canopy into sun-exposed and shaded leaves, m is a constant (0.68) based on the molecular weight ratio of CH_2O and CO_2, h accounts for the biochemical composition of plant biomass (carbohydrates, proteins, lipids), and p_a and p are atmospheric and leaf intercellular CO_2 partial pressure, respectively. For C_3 plants, (1- p_i/ p_a) approximately 0.3.

(Note: drainage and run-off are part of WU, but are not considered further here as plants only have a minor influence on them.)

TE at Field Level

Accordingly, yield or biomass production of a plant per unit of water use is expressed as transpiration efficiency (TE):

$$\text{Grain yield per water transpired } (TE_b) = B/T \quad (5)$$

$$\text{Biomass per water transpired } (TEg) = G/T \quad (6)$$

In Equation 5, either plant (aboveground) biomass at harvest or the incremental plant biomass production since the start of a treatment (usually withholding water) is used to determine B.

TE at Leaf Level

On a leaf level, carbon gain per water loss is defined as instantaneous transpiration efficiency (TE_{leaf}):

$$TE_{leaf} = AN/TR = (p_a \times (1 - p_i/p_a))/ (1.6 \times (e_i - e_a)) \quad (7)$$

Where, AN is net photosynthetic carbon assimilation and TR is the transpiration rate. Accounting for carbon losses via respiration (ϕc) and water losses via cuticular or night-time transpiration (ϕw; *i.e.* water losses that are not connected to CO_2 assimilation), the above equation is modified to (Hubick and Farquhar, 1989; Turner, 1997):

$$TE_{leaf} = (p_a \times (1 - p_i/p_a) \times (1 - \varphi c))/(1.6 \times (e_i - e_a)\times(1 - \varphi w)) \quad (8)$$

Most grain legumes are grown under rain-fed conditions, but, even in irrigated systems, yields tend to be constrained by available water (Daryanto *et al.*, 2015; Polania *et al.*, 2016). Moreover, because of their indeterminate nature and overlapping vegetative, reproductive and seed-filling stages, the critical period determining yield of grain legumes is longer than in cereals (Lake and Sadras, 2014). Extended periods of drought in low rainfall years, terminal drought at the end of the season, intermittent rainfall, and alternating periods of water deficit can reduce potential yields. This is particularly poignant in the light of predictions for many agro-ecological zones of more variable rainfall, and greater frequency and intensity of droughts under climate change (Wang, 2005; Zhao and Dai, 2016). To overcome these challenges, improving yield with less water, namely by increasing water use efficiency (WUE), is considered essential by many crop-breeding programmes (Araus *et al.*, 2002; Condon *et al.,* 2004; Araujo *et al.*, 2015). However, high water use efficiency (WUE) is generally associated with low yield potential (Blum, 2005), indicating that crops need to be both conservative and competitive in their water use (WU) in water-limited environments; that is, they need to be profligate water users with high transpiration rates (TRs) to maximize their productivity but restrict their water use (WU) so that it does not exceed supply. The transpiration coefficient (k) dependent function may not be applicable for water-limited conditions because it does not take into account stomatal or root sensitivity to soil water deficit and changes in the canopy as the crop develops. Furthermore, constant adequate soil water conditions are assumed for simplicity (Sinclair *et al*., 1984). Hence, a greater understanding of the mechanisms controlling water use (WU) and productivity of grain legumes is needed to develop crop ideotypes and better-adapted cultivars for different locations and farming systems. Integrating other properties into the water productivity framework is necessary. Matching these to well-characterized environments, including where high vapour pressure deficit (VPD) is applicable, will improve sustainability, reliability, and profitability of farming systems that incorporate grain legumes in their rotations.

Strategies of Water Use Relevant to The Environment

Despite similarities in total water use for the whole season, strategies of water use during the season can differ and indicate adaptation to different rainfall patterns. For example, grain legumes such as lupins can be both profligate and conservative in their water use depending on the susceptibility for terminal drought of their habitat (Berger and Ludwig, 2014). Profligate water use is a competitive strategy that enables the crop to take advantage of freely available water, where water loss to drainage, evaporation, neighbouring plants, or weeds is minimized. Yellow lupin (*Lupinus luteus*) and narrow-leafed lupin (L. *angustifolius*) roots are highly efficient at capturing and transporting water (Bramley *et al*., 2009), which supports their profligate water use. However, because lupins are predominantly grown and bred in Mediterranean-type environments prone to terminal drought (Berger and Ludwig, 2014), many genotypes are sensitive to soil drying, rapidly closing their stomata when soil moisture declines (Correia and Pereira, 1995). Stomatal closure may be necessary because they have limited capacity for osmotic adjustment (Palta *et al.*, 2012) and an inability to regulate root water transport reversibly (Bramley *et al.*, 2009, 2010). This prevents adaptation to drying conditions or to meeting the demands of transpiration. Moreover, lupins are not efficient in remobilizing resources from the stem to the sink, with $<3\%$ remobilized (Pate *et al.*, 1980). Thus, with little capacity to remobilize water-soluble carbohydrates, lupins rely on current photosynthetic carbon assimilation (AN) almost entirely to fill pods, which exacerbates yield reductions when they are exposed to terminal drought (Palta *et al.*, 2012). In a number of other species, stomatal conductance (gs) has been shown to respond to vapour pressure deficit ('atmospheric drought'), revealing a limited-traspiration trait, which has been reviewed elsewhere (Vadez *et al.*, 2012, 2014; Sinclair *et al.*, 2017). Longer water deficit conditions or environments characterized by stored soil moisture conditions may require predominantly low stomatal conductance and high transpiration efficiency (Moreno-Gutierrez *et al.*, 2012) to ensure that there is adequate water stored in the soil profile to be used by the crop during the critical reproductive phase. Restricted water use (i.e. low stomatal conductance and transpiration rate) early in the growing season as has been shown for chickpea (Zaman-Allah *et al.*, 2011b) helps to conserve soil water (Sinclair, 2012), ultimately resulting in higher water use at the reproductive stage Vadez *et al.* (2014). Post-flowering water use correlates with legume grain yields (Siddique *et al.*, 2001). Thus, conservative WU via early stomatal closure helps to reduce or delay terminal drought. Transpiration during the vegetative stage is often negatively related to that during the reproductive stage and yield (Ratnakumar *et al.*, 2009; Zaman-Allah *et al.*, 2011b), and could be used as a screening strategy in applicable environments, especially for legumes growing on stored soil moisture. A conservative strategy with temporal adjustment of water used over the growing season has the highest potential in systems with terminal drought.

Soil Evaporation: Influenced by Plant Habit & Agronomic Management

A. **Ground cover and Shading:** According to the agronomic definition of biomass per water used (WUE_b) (Equation1) or grain yield per water used (WUE_g) (Equation 2), soil evaporation is part of WU, thus soil evaporation negatively affects biomass per water used (WUE_b) or grain yield per water used (WUE_g). The level of ground cover and shading influence evaporative water losses from the soil, hence rapid establishment of ground cover (early vigour and prostrate growth habit), canopy architecture, and leaf orientation affect soil evaporation. For pea and faba bean, early ground cover results in lower soil evaporation and higher yields compared with chickpea or lentil, which tend to have low early vigour (Siddique *et al.*, 2001). In lupins, soil evaporation is inversely related to leaf area (Greenwood *et al.*, 1992), thus soil evaporation is a significant component of WU for plants with low leaf area. While early vigour and spreading/prostrate growth habit can potentially reduce evaporative water losses, high biomass accumulation and leaf area early in the season may lead to high Transpiration and exhaust stored soil moisture for later in the season (Ludlow and Muchow, 1990). Drought-tolerant chickpea genotypes showed lower vegetative growth under progressive soil drying (Zaman-Allah *et al.*, 2011a) and, even under well-watered conditions, drought-tolerant cowpea genotypes often have lower biomass compared with drought-sensitive genotypes (Belko *et al.*, 2013). When grown on stored soil moisture and low probability for in-season rainfall, as generally occurs in subtropical environments, early vigour might compromise yield of cool season grain legumes as early water consumption depletes the soil moisture profile and increases the severity of terminal drought. In contrast, in temperate and Mediterranean type cropping systems that rely on in-season rainfall, early ground cover prevents water losses through run-off and soil evaporation (Siddique *et al.*, 2012), and also provides a quicker canopy and root system to capture the available rainfall more readily Thus, a complex interaction between soil moisture and early vigour modifies the influence of plant biomass on soil evaporation, which could ultimately affect yield potential. Ground cover can be further modified by agronomic management techniques that adjust for plant morphology, such as row spacing, planting density, and stubble retention under reduced tillage systems. For example, chickpea cultivars planted into standing wheat stubble and faba bean cultivars had higher yields and higher grain yield per water used (WUE_g) under narrower row spacings (Raymond *et al.*, 2015).

B. **Stubble Retention:** Stubble retention (particularly stubble from cereals in rotation) in reduced tillage systems helps to reduce soil evaporation and thus increase the water available for the following grain legume crops (Kirkegaard and Ryan, 2014). Stubble, and whether it is left standing or flattened, can also affect infiltration of rainfall, and soil and canopy temperatures of following grain legumes such as chickpea and faba bean (Verrell, 2016). Standing wheat residue reduced virus infections and incidence of radiant frost in chickpea (Verrell, 2016). Cooler canopy temperatures in reduced tillage systems also means less water is needed for the cooling effect of transpiration. Effects of reduced tillage systems on yield or water use efficiency of grain legumes are sometimes conflicting. For example, zero tillage had no impact on WUE_g of chickpea (Fernandez-Garcia *et al.*, 2013) or yield of faba bean (Lopez-Bellido *et al.*, 2003), whereas WUE_g was higher in chickpea in a tilled system because the preceding fallow season allowed more deep water storage in the soil profile (Gan *et al,* 2010). Conservation agriculture and reduced tillage are widely adopted, especially in countries such as Australia. A greater understanding of the interaction between stubble retention, species/genotype, and the site-specific climate and soil conditions is needed to exploit its full potential to improve yields of grain legumes. Wild grain legumes that are the progenitors of domesticated species tend to have low stature and prostrate growth or climbing habit (Kelly, 2001; Abbo *et al.*, 2009). These adaptations aid light interception and water conservation in the environments in which these species originate, such as higher altitudes and shallow soils, but they are not conducive to high-density planting, crop management, and harvesting by agricultural machinery (Kumar *et al.*, 2024). Nevertheless, the wild material provides opportunities for adding novel traits that aid water conservation.

Phenological Adjustment to Escape Drought

Escaping drought through early completion of the full plant cycle subject to terminal drought is well known for annual grain legumes (Vadez *et al.*, 2012; Berger *et al.*, 2016). In agricultural systems, this is achieved by early planting (provided there are no restrictions through frost, photoperiod, or waterlogging), and choice of early flowering and maturing (short growing season) species and genotypes (Siddique *et al.*, 2001). These agronomic practices and traits reduce impairments in yield through terminal drought. Escaping drought becomes extremely prominent when grain legumes grow on stored soil moisture and severe terminal drought is inevitable-then drought avoidance traits become ineffective (Berger *et al.*, 2016). As discussed above, early accumulation of biomass also contributes to prevent water losses via soil evaporation. In wet seasons, however, early maturity and a shorter growth cycle counteract temporal accumulation of carbon (higher yield potential, as has been observed for lentil (Silim *et al.*, 1993). Early vigour and flowering were also associated with larger seeds, which was supported by a longer pod-filling phase (Berger *et al.*, 2017).

Physiological Adjustments to Environmental Conditions

A. **Root systems:** Root traits such as rooting depth, branching, and anatomy influence water uptake, root hydraulic resistance, and plant water balance. Root hydraulic resistance is further modified by the presence and activity of aquaporins (proteinaceous water channels) facilitating transcellular water transport driven by osmotic and hydrostatic pressure gradients (Tyerman *et al.*, 2002). While no individual root trait is primarily responsible for crop performance or drought tolerance in the field, genetic variations in several traits have been identified in grain legume roots that enhance water transport and, hence, the ability to match Transpiration demands and maintain shoot water balance. The utility of root length density (RLD; total root length per unit volume of soil) as a trait to improve yield under water-sufficient and water-deficient conditions has been suggested (Brodribb *et al.*, 2015) because water uptake by roots depends on the total absorptive root surface area. Hence, root systems with high RLD should have high hydraulic conductance (inverse of hydraulic resistance) and thus increase the amount of transpirable water (Vadez, 2014). In support of this, a positive relationship between total RLD and grain yield was observed for chickpea (Kashiwagi *et al.*, 2006a; Purushothaman *et al.,* 2017), and drought tolerance of cowpea has been attributed to increased root biomass and higher root length density (RLD) at greater depth (Matsui and Singh, 2003). Typically, root length density (RLD) is higher in shallower soils and decreases with increasing depth (Hamblin and Tennant, 1987), reflecting the distribution of available nutrients. Thus, water uptake is generally greater in shallower soils (Hamblin and Tennant, 1987).

B. **Osmotic adjustment:** Osmotic adjustment is the increased solute concentration in cells through the active accumulation of ions and compatible solutes, which aids turgor maintenance, and has recently been reviewed and discussed (Blum, 2017; Turner, 2017). Although osmotic adjustment has been suggested as a useful selection criterion associated with grain yield under drought (Morgan *et al.*, 1991), the results for chickpea indicate that it is not a stable trait (Basu *et al.*, 2007) and does not result in yield improvement (Turner *et al.*, 2007a). Moreover, osmotic adjustment may not be a character shared by every grain legume; for example, lupins apparently do not osmotically adjust (Palta *et al.*, 2012). However, in a meta-analysis of 26 studies across 12 crop species, which included four grain legumes-chickpea, soybean, pigeon pea, and pea yield under drought was associated with osmotic adjustment in almost all cases, including the grain legumes (Blum, 2017). Besides osmotic adjustment, accumulation of metabolites such as proline has also been related to osmo-protection of cellular functions in cowpea (Goufo *et al.*, 2017).

C. **Stomatal conductance (gs):** Stomata open and close in response to environmental drivers, such as light, CO_2 concentration and vapour pressure deficit (VPD). Because stomata are the major pathway for gas exchange between plants and the atmosphere, net photosynthetic carbon assimilation (AN) and transpiration (T) are intrinsically tied. However, T increases linearly with increasing stomatal conductance, whereas net photosynthetic carbon assimilation (AN) reaches a plateau at higher gs (Yoo *et al.*, 2009). Generally, low stomatal conductance helps to conserve soil water. Under conditions with no water stress (soil water deficit or atmospheric drought), reduced stomatal conductance results in lower yield, while soil water conservation due to lower stomatal conductance is beneficial under terminal drought or high vapour pressure deficit conditions. Besides the direct effect on stomatal conductance, vapour pressure deficit also drives transpiration efficiency at leaf level (at high vapour pressure deficit relatively more water is lost per assimilated carbon, thus lower transpiration efficiency at leaf level; Equations 7 and 8). Therefore, it is favourable for water conservation to limit stomatal conductance/ transpiration rate and thus increase transpiration efficiency at leaf level (TE_{leaf}) under conditions with high vapour pressure deficit; a trait that has been associated with drought tolerance and was recently reviewed by Sinclair *et al.* (2017).

Other Ways to Reduce Water Losses and Increase Carbon Gain

A. **Re-fixation of respired CO_2:** The presence of chlorophyll in the embryo and seed coat (Furbank *et al.*, 2004) characterizes legumes as chloro-embryophytes (Smolikova and Medvedev, 2016). The net carbon assimilation of pods is small (Ma *et al.*, 2001) and does not substantially contribute to the overall plant carbon gain. However, pod wall photosynthesis measurably contributes to seed filling in the perennial forage legume alfalfa, being especially important when leaves start to senescence (Wang *et al.*, 2016). As respired CO_2 is trapped in the cavity of the pod, it can be re-assimilated at no expense of water loss. Re-fixation of respired CO_2 by chickpea increases gross CO_2 assimilation by the pods (Ma *et al.*, 2001), but it is the pod wall and not the seed that does most of the re-fixation (Furbank *et al.*, 2004). The chlorophyll content in pod walls has even been shown to increase early in pod development due to water deficit in a deep-rooted chickpea genotype (Kaur *et al.*, 2017), thus highlighting the potential importance of re-fixation of respired CO_2 under drought, especially for species that lose their leaves prior to terminal drought. The contribution of CO_2 re- fixation might be higher in top-podding grain legumes, where, for example, flowers and pods of some genotypes of mungbean (*Vigna radiata*), pigeon pea, and cowpea are able to intercept a larger proportion of light (Lawn, 1989). Further research is needed to compare the capacity of re-fixation of respired CO_2 of different grain legumes and to quantify its contribution under water deficit.

B. Effective leaf area: Besides the transpiration rate (TR), transpirational water losses are determined by the leaf area. However, reduced leaf area due to limited leaf expansion or leaf senescence concurrently limits photosynthetic carbon uptake and yield potential (Blum, 2005) and does not, therefore, result in better water use efficiency/ transpiration efficiency. Under conditions of terminal drought, plants benefit from smaller leaf area early in the season as that contributes to soil water conservation by reduced transpirational water losses (Ludlow and Muchow, 1990).

C. N_2 fixation: Besides assimilation of soil NH_4 and NO_3 , legumes have the ability to source nitrogen via biological N_2 fixation. Because legumes provide carbon to symbiotic rhizobia bacteria, biological N_2 fixation represents a carbon sink due to the supply of energy for nitrogenase and carbon skeletons to fix ammonia. The carbon cost associated with N_2 fixation is generally higher compared with soil nitrogen assimilation (Cannell and Thornley, 2000). N_2-fixing plants were less sensitive to water deficit compared with nitrate-fed plants in terms of plant growth (pea; Frechilla *et al*., 2000), maintaining water status (soybean; Kirova *et al*., 2008) and pod yield (common bean; Lodeiro *et al*., 2000). Consequently, the source of N affects both factors of transpiration efficiency at leaf level (TE_{leaf}) under water deficit, so their interaction determines the overall effect on transpiration efficiency (TE). Symbiotic N_2 fixation could mediate increased photosynthetic capacity due to increased leaf N concentrations, thus allowing high transpiration efficiency at leaf level (TE_{leaf}) without sacrificing yield potential (Condon *et al.,* 2002) a unique potential for legumes. Due to those potential and contrasting results on yield benefits under water deficit conditions, more research is needed on the effect of N_2 fixation on transpiration efficiency (TE) and water use efficiency (WUE).

Nitrogen Transfer Mechanism

The major routes of N transfer can be categorized as aboveground and belowground(Rouquette and Smith, 2010; Peoples *et al*., 2015). Belowground N transfer, involves three distinct pathways that is sown in (Fig. 6.1).

Decomposition: Decomposition of legume root tissues followed by uptake of released N by neighouring plants (Fustec *et al.*, 2010).

Root exudation: The exudation of soluble N compounds by donors and uptake by receivers (Gylfadottir *et al.*, 2007; Paynel *et al.*, 2008; Wichern *et al.*, 2008).

Mycorrhizae: Transfer of N mediated by plant-associated mycorrhizae (He *et al.*, 2003, 2009).

Nitrogen transfer is often described as either "direct" or "indirect," but not all of these processes fit nicely into a single category. Direct N transfer is the transfer of N from a donor plant to receiver plant without undergoing mineralization, whereas indirect N transfer involves mineralization followed by uptake of N compounds by the receiver. The common mycelial network, which interconnects the root systems of nearby plants (Haystead *et al.*, 1988; Dubach and Russelle, 1994; He *et al.*, 2003), is the most direct path for the transfer of N from donor to receiver (Fig. 6.1). However, mycorrhizae also facilitate indirect N transfer through uptake and transfer of N derived from decomposing organic matter of the donor (e.g., legume roots and nodule debris) (Fig. 1); this can only occur after the root system of the receiver plant is colonized with mycorrhizae (Leigh *et al.,* 2009). The root exudate pathway has been considered both direct (Paynel *et al.*, 2008) and indirect (Jalonen *et al.*, 2009a), depending on whether the N exudate is taken up immediately by neighbouring plants (direct N transfer) or initially mineralized by soil microbiota (indirect N transfer). The decomposition pathway can be considered to be the most indirect, as N from root debris must be first decomposed and mineralized before it is accessible for uptake by receiver roots (Fustec *et al.*, 2010).

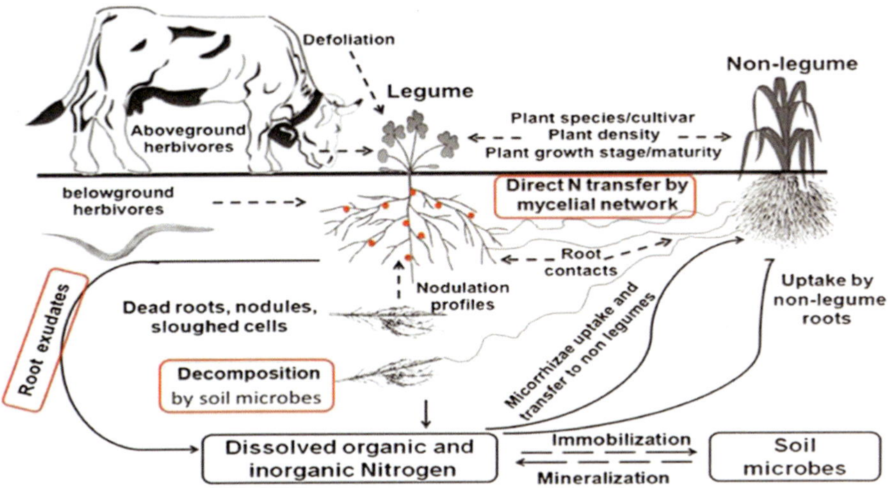

Fig. 6.1: Potential mechanisms of below-ground nitrogen transfers from legumes to non-legumes (decomposition of nodules and roots, root exudates, and mycorrhizal-mediated nitrogen transfer) and the wide range of abiotic and biotic aspects that influence nutrient transfers

Source: Thilakarathna *et al.* (2016)

Summary and Conclusion

The effects of traits contributing to yield of legumes, highly depend on the regime of water; that is, no water limitation, terminal drought, or transient drought. Under conditions without water limitations or transient drought, a profligate water use generally helps to increase yields, while under terminal drought a more conservative water use improves yields through soil water conservation in combination with adapted phenology to escape terminal drought. Multiple traits exhibit contrasting effects under conditions without water limitation and terminal drought because traits that help to reduce early soil moisture depletion, such as reduced stomatal conductance, late vigour, or reduced leaf area, compromise yield potential. Under terminal drought or on soils with low water-holding capacity, any traits to avoid/postpone drought become ineffective.

Therefore, it is crucial to characterize the environment (i.e. typical climate conditions, likelihood for extreme weather events, and soil properties) to match with the best adapted species and genotypes. The traits discussed above can be classified into constitutive (not reversible, e.g. root length) and adjustable/reversible (e.g. stomatal adjustment to vapour pressure deficit) traits. Generally, reversible traits and phenotypic plasticity are preferred in order to maintain high yield potential; thus, for example, stomatal closure at high vapour pressure deficit rather than constantly low stomatal conductance.

However, the suitability of constitutive versus reversible traits also depends on the likelihood and regime of drought (terminal versus transient). Water use and Water use efficiency at field level depend on multiple phenological, morphological, and physiological traits, as well as on agronomic practices. Because all these attributes converge into the final grain legume yield, interdisciplinary approaches are essential, which should also involve a social component in order to translate breeding and agronomic achievements into economic outcome for farmers.

Acknowledgement

All the authors are thankful to Bihar Agricultural University, Sabour, Bihar, India via BAU Communication No. - 1829/240821.

References

Abbo, S., Saranga, Y., Peleg, Z., Kerem, Z., Lev-Yadun, S., & Gopher, A. (2009). Reconsidering domestication of legumes versus cereals in the ancient near east. Quarterly Review of Biology, 84, 29–50.

Ali, M. (2008). Technological Approaches for increasing pulse production in India. In Proceedings of National Symposium on Technological Innovation for Resource Starved Farmers in Global Perspective, CSAUA & T Kanpur 28-30 April.

Ali, M., & Gupta, S. (2012). Carrying capacity of Indian agriculture: pulse crops. Cur. Sci., 102, 874- 881.

Anderson, W., Johansen, C and Siddique, K.H.M. (2016). Addressing the yield gap in rainfed crops: a review. Agronomy for Sustainable Development, 36, 18.

Angadi, S.V., McConkey, B.G., Cutforth, H.W., Miller, P.R., Ulrich, D., Selles, F., Volkmar, K.M., Entz, M.H., & Brandt, S.A. (2008). Adaptation of alternative pulse and oilseed crops to the semiarid Canadian Prairie: seed yield and water use efficiency. Canadian Journal of Plant Science, 88, 425–438.

Araujo, S.S., Beebe, S., & Crespi, M. (2015). Abiotic stress responses in legumes: strategies used to cope with environmental challenges. Critical Reviews in Plant Sciences, 34, 237–280.

Araus, J.L., Slafer, G.A., Reynolds, M.P., & Royo, C. (2002). Plant breeding and drought in C3 cereals: what should we breed for? Annals of Botany, 89, 925–940.

Bampidis, V.A., & Christodoulou, V. (2011). Chickpeas (Cicer arietinum L.) in animal nutrition: a review. Animal Feed Science and Technology, 168, 1–20.

Basu, P.S., Berger, J.D., Turner, N.C, Chaturvedi, S.K, Ali, M., & Siddique, K.H.M. (2007). Osmotic adjustment of chickpea (Cicer arietinum) is not associated with changes in carbohydrate composition or leaf gas exchange under drought. Annals of Applied Biology, 150, 217–225.

Belda, M., Holtanova, E., Halenka, T., & Kalvová, J. (2014). Climate classification revisited: from Koppen to Trewartha. Climate Research, 59, 1–13.

Belko, N., Zaman-Allah, M., Diop, N.N, Cisse, N., Zombre, G., Ehlers, J.D., & Vadez, V. (2013). Restriction of transpiration rate under high vapour pressure deficit and non limiting water conditions is important for terminal drought tolerance in cowpea. Plant Biology, 15, 304–316.

Berger, J., Palta, J., & Vadez, V. (2016). Review: an integrated framework for crop adaptation to dry environments: responses to transient and terminal drought. Plant Science, 253, 58–67.

Berger, J.D., & Ludwig, C. (2014). Contrasting adaptive strategies to terminal drought stress gradients in Mediterranean legumes: phenology, productivity, and water relations in wild and domesticated Lupinus luteus L. Journal of Experimental Botany, 65, 6219–6229.

Berger, J.D., Shrestha, D., & Ludwig, C. (2017). Reproductive strategies in Mediterranean legumes: trade-offs between phenology, seed size and vigor within and between wild and domesticated lupinus species collected along aridity gradients. Frontiers in Plant Science, 8, 548.

Blum, A. (2005). Drought resistance, water-use efficiency and yield potential—are they compatible, dissonant, or mutually exclusive? Australian Journal of Agricultural Research, 56, 1159–1168.

Blum, A. (2017). Osmotic adjustment is a prime drought stress adaptive engine in support of plant production. Plant, Cell and Environment, 40, 4–10.

Bramley, H., Turner, N.C., Turner, D.W., & Tyerman, S.D. (2010). The contrasting influence of short-term hypoxia on the hydraulic properties of cells and roots of wheat and lupin. Functional Plant Biology, 37, 183–193.

Bramley, H., Turner, N.C., Turner, D.W., & Tyerman, S.D. (2009). Roles of morphology, anatomy, and aquaporins in determining contrasting hydraulic behavior of roots. Plant Physiology, 150, 348–364.

Brodribb, T.J., Holloway-Phillips, M.M., & Bramley, H. (2015). Improving water transport for carbon gain in crops. In: Calderini DF, Sadras VO, eds. Crop physiology, 2nd edn. San Diego: Academic Press, pp. 251–281.

Cannell, M.G.R., & Thornley, J.H.M.(2000). Modelling the components of plant respiration: some guiding principles. Annals of Botany, 85, 45–54.

Cernay, C., Ben-Ari, T., Pelzer, E., Meynard, J.M., & Makowski, D. (2015). Estimating variability in grain legume yields across Europe and the Americas. Scientific Reports, 5, 11171.

Cernay, C., Pelzer, E., & Makowski, D. (2016). A global experimental dataset for assessing grain legume production. Scientific Data, 3, 160084.

Correia, M.J., & Pereira, J.S. (1995). The control of leaf conductance of white lupin by xylem ABA concentration decreases with the severity of water deficits. Journal of Experimental Botany, 46, 101–110.

Daryanto, S., Wang, L., & Jacinthe, P.A. (2015). Global synthesis of drought effects on food legume production. PLoS One, 10, e0127401.

Daryanto, S., Wang, L., & Jacinthe, P.A. (2017). Global synthesis of drought effects on cereal, legume, tuber and root crops production: a review. Agricultural Water Management, 179, 18–33.

Dubach, M., & Russelle, M.P. (1994) Forage legume roots and nodules and their role in nitrogen transfer. Agron Journal, 86, 259–266.

Fernandez-Garcia,P., Lopez-Bellido, L., Munoz-Romero, V., & Lopez-Bellido, R.J. (2013). Chickpea water use efficiency as affected by tillage in rainfed Mediterranean conditions. Agriculural Managemnent Systems, 129, 194–199.

Foyer, C.H., Lam, H.M., & Nguyen, H.T. (2016). Neglecting legumes has compromised human health and sustainable food production. Nature Plants, 2, 16112.

Frechilla, S., Gonzalez, E.M., Royuela, M., Minchin, F.R., Aparicio-Tejo, P.M., & Arrese-Igor C. (2000). Source of nitrogen nutrition (nitrogen fixation or nitrate assimilation) is a major factor involved in pea response to moderate water stress. Journal of Plant Physiology, 157, 609–617.

French, R.J., & Schultz, J.E. (1984). Water use efficiency of wheat in a mediterranean type environment.The relation between yield, water use and climate. Australian Journal of Agricultural Research, 35, 743–764.

Furbank, R.T., White, R., Palta, J.A., & Turner, N.C. (2004). Internal recycling of respiratory CO_2 in pods of chickpea (Cicer arietinum L.): the role of pod wall, seed coat, and embryo. Journal of Experimental Botany, 55, 1687–1696.

Fustec, J., Lesuffleur, F., Mahieu, S., & Cliquet, J.B. (2010). Nitrogen rhizo deposition of legumes. A review. Agron Sustain Dev., 30, 57–66.

Gan, Y.T., Warkentin, T.D., Bing, D.J., Stevenson, F.C., & Mc Donald, C.L. (2010). Chickpea water use efficiency in relation to cropping system, cultivar, soil nitrogen and rhizobial inoculation in semiarid environments. Agricultural Water Management, 97, 1375–1381.

Greenwood, E.A.N., Turner, N.C., Schulze, E.D., Watson, G.D., & Venn, N.R. (1992). Groundwater management through increased water use by lupin crops. Journal of Hydrology, 134, 1–11.

Gylfadottir, T., Helgadottir, A., & Hogh-Jensen, H. (2007). Consequences of including adapted white clover in northern European grassland: transfer and deposition of nitrogen. Plant Soil, 297, 93–104.

Hamblin, A., & Tennant, D. (1987). Root length density and water uptake in cereals and grain legumes: how well are they correlated? Australian Journal of Agricultural Research, 38, 513–527.

Hauggaard-Nielsen, H., Jornsgaard, B., Kinane, J., & Jensen, E.S. (2008). Grain legume– cereal intercropping: the practical application of diversity, competition and facilitation in arable and organic cropping systems. Renewable Agriculture and Food Systems, 23, 3–12.

Haystead, A., Malajczuk, N., & Grove, T. (1988) Underground transfer of nitrogen between pasture plants infected with vesicular-arbuscular mycorrhizal fungi. New Phytol., 108, 417–423.

He, X.H., Critchley, C., & Bledsoe, C. (2003). Nitrogen transfer within and between plants through common mycorrhizal networks (CMNs). Crit Rev Plant Sci., 22, 531–567.

Hubick, K., & Farquhar, G. (1989). Carbon isotope discrimination and the ratio of carbon gained to water lost in barley cultivars. Plant, Cell and Environment, 12, 795–804.

Jalonen, R., Nygren, P., & Sierra, J. (2009a). Transfer of nitrogen from a tropical legume tree to an associated fodder grass via root exudation and common mycelial networks. Plant Cell Environ., 32, 1366–1376.

Kashiwagi, J., Krishnamurthy, L., Crouch, J.H., & Serraj, R. (2006a). Variability of root length density and its contributions to seed yield in chickpea (Cicer arietinum L.) under terminal drought stress. Field Crops Research, 95, 171–181.

Kaur, D., Grewal, S.K., Kaur, J., & Singh, S. (2017). Free radical scavenging activities can mitigate the effect of water stress in chickpea. Crop and Pasture Science, 544–554.

Keller, A., Sakthivadivel, R., & Seckler, D. (2000). 'Water scarcity and the role of storage in development' IWMI research report 39, IWMI: Colombo.

Kelly, J.D. (2001). Remaking bean plant architecture for efficient production. Advances in Agronomy, 71, 109–143.

Kirkegaard, J.A., & Ryan, M.H. (2014). Magnitude and mechanisms of persistent crop sequence effects on wheat. Field Crops Research, 164, 154–165.

Kirova, E., Tzvetkova, N., Vaseva, I., & Ignatov, G. (2008). Photosynthetic responses of nitrate-fed and nitrogen-fixing soybeans to progressive water stress. Journal of Plant Nutrition, 31, 445–458.

Kottek, M., Grieser, J., Beck, C., Rudolf, B., & Rubel, F. (2006). World map of the Koppen–Geiger climate classification updated. Meteorologische Zeitschrift, 15, 259–263.

Kumar, N., Sow, S., Rana, L., Kumar, V., Kumar, J., Pramanick, B., et al. (2024). Productivity, water use efficiency and soil properties of sugarcane as influenced by trash mulching and irrigation regimes under different planting systems in sandy loam soils. Frontiers in Sustainable Food Systems, 8, 1340551.

Lake, L., & Sadras, V.O. (2014). The critical period for yield determination in chickpea (Cicer arietinum L.). Field Crops Research, 168, 1–7.

Lawn, R.J. (1989). Agronomic and physiological constraints to the productivity of tropical gain legumes and prospects for improvement. Experimental Agriculture, 25, 509–528.

Leigh, J., Hodge, A., & Fitter, A.H. (2009) Arbuscular mycorrhizal fungi can transfer substantial amounts of nitrogen to their host plant from organic material. New Phytol., 181, 199–207.

Lenka, N.K., & Lal, R. (2012). Soil related constraints to the CO_2 fertilization effect. Crit. Rev. Plant Sci., 31, 342–357.

Lodeiro, A.R., Gonzalez, P., Hernandez, A., Balague, L.J., & Favelukes, G. (2000). Comparison of drought tolerance in nitrogen-fixing and inorganic nitrogen grown common beans. Plant Science, 154, 31–41.

Lopez-Bellido, R.J., Lopez-Bellido, L., Lopez-Bellido, F., & Castillo, J.E. (2003). Faba bean (Vicia faba L.) response to tillage and soil residual nitrogen in a continuous rotation with wheat (Triticum aestivum L.) under rainfed Mediterranean conditions. Agronomy Journal, 95, 1253–1261.

Ludlow, M.M., & Muchow, R.C. (1990). A critical evaluation of traits for improving crop yields in water-limited environments. Advances in Agronomy, 43, 107–153.

Ma, Q., Behboudian, M.H., Turner, N.C., & Palta, J.A. (2001). Gas exchange by pods and subtending leaves and internal recycling of CO_2 by pods of chickpea (Cicer arietinum L.) subjected to water deficits. Journal of Experimental Botany, 52, 123–131.

Maphosa, Y., & Jideani, V.A. (2017). The role of legumes in human nutrition in. In: Hueda, M.C. (Ed.), Functional Food-Improve Health through Adequate Food. Intech open publishing, Croatia, p. 13.

Matsui, T and Singh, B.B. (2003). Root characteristics in cowpea related to drought tolerance at the seedling stage. Experimental Agriculture, 39, 29–38.

Millan, T., Clarke, H.J., Siddique, K.H., Buhariwalla, H.K., Gaur, P.M., Kumar, J., Gil, J., Kahl, G., & Winter, P. (2006). Chickpea molecular breeding: new tools and concepts. Euphytica, 147, 81–103.

Moreno-Gutiérrez, C., Dawson, T.E., Nicolas, E., & Querejeta, J.I. (2012). Isotopes reveal contrasting water use strategies among coexisting plant species in a Mediterranean ecosystem. New Phytologist, 196, 489–496.

Morgan, J.M., Rodriguez-Maribona, B., & Knights, E.J. (1991). Adaptation to water deficit in chickpea breeding lines by osmoregulation: relationship to grain yields in the field. Field Crops Research, 27, 61–70.

Palta, J.A., Berger, J.D., & Bramley, H. (2012). Physiology of the yield under drought: lessons from studies with lupin. In: Aroca R, ed. Plant responses to drought stress. Berlin: Springer, 417–440.

Passioura, J.B. (1977). Grain yield, harvest index, and water use of wheat. Journal of the Australian Institute of Agricultural Science, 43, 117–120.

Pate, J.S., Atkins, C.A., & Perry, M.W. (1980). Significance of photosynthate produced at different stages of growth as carbon source for fruit filling and seed reserve accumulation in Lupinus angustifolius L. Australian Journal of Plant Physiology, 7, 283.

Paynel, F., Lesuffleur, F., & Bigot, J. (2008) A study of ^{15}N transfer between legumes and grasses. Agron Sustain Dev., 28, 281–290.

Peoples, M.B., Chalk, P.M., Unkovich, M.J., & Boddey, R.M. (2015). Can differences in ^{15}N natural abundance be used to quantify the transfer of nitrogen from legumes to neighbouring non-legume plant species? Soil Biol Biochem, 87, 97–109.

Polania, J.A., Poschenrieder, C., Beebe, S., & Rao, I.M. (2016). Effective use of water and increased dry matter partitioned to grain contribute to yield of common bean improved for drought resistance. Frontiers in Plant Science, 7, 660.

Purushothaman, R., Krishnamurthy, L., Upadhyaya, H.D., Vadez, V., & Varshney, R.K. (2017). Genotypic variation in soil water use and root distribution and their implications for drought tolerance in chickpea. Functional Plant Biology, 44, 235–252.

Ratnakumar, P., Vadez, V., Nigam, S.N., & Krishnamurthy, L.(2009). Assessment of transpiration efficiency in peanut (Arachis hypogaea L.) under drought using a lysimetric system. Plant Biology, 11 (Suppl 1), 124–130.

Raymond, R., McKenzie, K., & Rachaputi, R.C.N. (2015). Impact of row spacing on chickpea, faba bean and mung bean. GRDC updates. Rome: GRDC Grains Research.

Rosenzweig, C., & Hillel, D. (1995). Potential impacts of climate change on agriculture and food supply. Consequences, 1, 23–32.

Rouquette, F.M., & Smith, G.R. (2010) Review: effects of biological nitrogen fixation and nutrient cycling on stocking strategies for cow-calf and stocker programs. Prof Anim Sci., 26, 131–141

Siddique, K.H.M., Johansen, C., Turner, N.C., Jeuffroy, M.H., Hashem, A., Sakar, D., Gan, Y., & Alghamdi, S.S. (2012). Innovations in agronomy for food legumes. A review. Agronomy for Sustainable Development, 32, 45–67.

Siddique, K.H.M., Regan, K.L., Tennant, D., & Thomson, B.D. (2001). Water use and water use efficiency of cool season grain legumes in low rainfall Mediterranean-type environments. European Journal of Agronomy, 15, 267–280.

Silim, S.N., Saxena, M.C., & Erskine, W. (1993). Adaptation of lentil to the mediterranean environment. Factors affecting yield under drought conditions. Experimental Agriculture, 29, 9–19.

Sinclair, T.R. (2012). Is transpiration efficiency a viable plant trait in breeding for crop improvement? Functional Plant Biology, 39, 359–365.

Sinclair, T.R., Devi, J., Shekoofa, A., Choudhary, S., Sadok, W., Vadez, V., Riar, M., & Rufty, T. (2017). Limited-transpiration response to high vapor pressure deficit in crop species. Plant Science, 260, 109–118.

Sinclair, T.R., Tanner, C.B., & Bennett, J.M. (1984). Water-use efficiency in crop production. Bio Science, 34, 36–40.

Sinha, S.K. (1977). Food legumes: distribution, adaptability and biology of yield. Rome: FAO.

Smolikova, G.N. and Medvedev, S.S. (2016). Photosynthesis in the seeds of chloroembryophytes. Russian Journal of Plant Physiology, 63, 1–12.

Thilakarathna, M. S., McElroy, M. S., Chapagain, T., Papadopoulos, Y. A., & Raizada, M. N. (2016). Belowground nitrogen transfer from legumes tonon-legumes under managed herbaceous cropping systems. A review. Agron.Sustain. Dev., p. 368

Turner, N.C. (1997). Further progress in crop water relations. In: Sparks DL ed. Advances in Agronomy, 58, 293–338.

Turner, N.C. (2017). Turgor maintenance by osmotic adjustment, an adaptive mechanism for coping with plant water deficits. Plant Cell and Environment, 40, 1–3.

Turner, N.C., Abbo, S., Berger, J.D., Chaturvedi, S.K., French, R.J., Ludwig, C., Mannur, D.M., Singh, S.J., & Yadava, H.S. (2007a). Osmotic adjustment in chickpea (Cicer arietinum L.) results in no yield benefit under terminal drought. Journal of Experimental Botany, 58, 187–194.

Tyerman, S.D., Niemietz, C.M., & Bramley, H. (2002). Plant aquaporins: multifunctional water and solute channels with expanding roles. Plant Cell and Environment, 25, 173–194.

Vadez, V. (2014). Root hydraulics: the forgotten side of roots in drought adaptation. Field Crops Research, 165, 15–24.

Vadez, V., Berger, J.D., & Warkentin, T. (2012). Adaptation of grain legumes to climate change: a review. Agronomy for Sustainable Development, 32, 31–44.

Vadez, V., Kholova, J., Medina, S., Kakkera, A., & Anderberg, H. (2014). Transpiration efficiency: new insights into an old story. Journal of Experimental Botany, 65, 6141–6153.

Verrell, A. (2016). Stubble and its impact on temperature in chickpea crops. GRDC update papers. Rome: GRDC Grains Research. Google Scholar Google Preview World.

Wang, G. (2005). Agricultural drought in a future climate: results from 15 global climate models participating in the IPCC 4th assessment. Climate Dynamics, 25, 739–753.

Wang, N., Hatcher, D.W., Toews, R., & Gawalko, E.J., (2009). Influence of cooking and dehulling on nutritional composition of several varieties of lentils (Lens culinaris). LWT Food Sci. Technol., 42, 842–848.

Wichern, F., Eberhardt, E., & Mayer, J. (2008). Nitrogen rhizodeposition in agricultural crops: methods, estimates and future prospects. Soil Biol Biochem, 40, 30–48.

Yoo, C.Y., Pence, H.E., Hasegawa, P.M., & Mickelbart, M.V. (2009). Regulation of transpiration to improve crop water use. Critical Reviews in Plant Sciences, 28, 410–431.

Zaman-Allah, M., Jenkinson, D.M., & Vadez, V. (2011a). Chickpea genotypes contrasting for seed yield under terminal drought stress in the field differ for traits related to the control of water use. Functional Plant Biology, 38, 270–281.

Zaman-Allah, M., Jenkinson, D.M., & Vadez, V. (2011b). A conservative pattern of water use, rather than deep or profuse rooting, is critical for the terminal drought tolerance of chickpea. Journal of Experimental Botany, 62, 4239–4252.

Zhao, T., & Dai, A. (2016). Uncertainties in historical changes and future projections of drought. Part II: model-simulated historical and future drought changes. Climatic Change, 144, 535–549.

7

Genetic Diversity, Conservation and Improvement of Grain Legumes

Kumari Rashmi[1*], Sanjay Sahay[2], Ravi S. Singh[1], Anushree[1], Adyant Kumar[3], Mahesh Kumar[4] and Vikash Kumar[2]

[1]Department of Plant Breeding and Genetics, DKAC, Kishanganj, BAU, Bihar, India

[2]Department of Horticulture (Fruit & Fruit Technology), DKAC, Kishanganj, BAU Bihar, India

[3]Department of Agronomy, DKAC, Kishanganj, BAU, Bihar, India

[4]Department of Molecular Biology & Genetic Engineering, DKAC, Kishanganj, BAU Bihar, India

**Corresponding author's email: versatilerash@rediffmail.com*

Abstract

Understanding genetic diversity is essential for achieving genetic improvement and conservation of grain legumes. These crops serve as important pulses grown and consumed all over the world, especially in Africa, South East Asia and America. Legumes serve as a good source of carbohydrates, oil, fibre and proteins. Grain legumes are a major source of food and nutrition globally. However, it has not been possible to enhance gain legume production to the required level, primarily due to the narrow genetic base of most of the legume crops, coupled with changing climatic conditions. Most grain legumes are lacking in desired plant ideotypes and resistance sources for various biotic and abiotic stresses. The genetic diversity conserved in gene banks globally is a major resource for crop breeding programs, but it is utilized only marginally. Therefore, in order to broaden the genetic base of grain legume crops and enhance the genetic gains made in improvement programs, conventional approaches and modern scientific tools should be integrated in a phased and carefully judged manner. The first phase should focus on the search for desired traits and the infusion of diversity into the cultivated gene pool through use of landraces and CWRs. The second phase should focus on the utilization of advanced selection tools, such as genomics, high-throughput precision phenotyping, and artificial intelligence, to exploit the hidden potential of the available genetic diversity; and the third phase should involve technologies such as mutational breeding, genome editing, and transgenic technologies to improve, modify any novel or alien traits that are not available in the entire crop gene pool. We presume that enrichment of the genetic diversity of cultivated grain legume gene pools, along with simultaneous improvements in their yield and plant type with the aid of advanced scientific tools, will enhance grain legume crop yields to the required level.

Keywords**:** *Crop improvement, Genetic diversity, Genome editing, Legume, Phenotyping*

Introduction

Economically, legumes (Fabaceae) represent the second most important family of crop plants after the grass family, Poaceae. Grain legumes account for 27% of world crop production and provide 33% of the dietary protein consumed by humans, while pasture and forage legumes provide vital part of animal feed. Fabaceae, the third largest family of flowering plants, has traditionally been divided into the following three subfamilies: Caesalpinioideae, Mimosoideae, and Papilionoideae, all together with 800 genera and 20,000 species. The latter subfamily contains most of the major cultivated food and feed crops. Among the grain legumes are some of mankind's earliest crop plants, whose domestication parallelled that of cereals: Soybean in China; faba bean, lentil, chickpea and pea in the Fertile Crescent of the Near East; cowpeas and bambara groundnut in Africa; soybean and mungbeans in East Asia; pigeonpea and the grams in South Asia; and common bean, lima bean, scarlet runner bean, tepary bean and lupin in Central and South America. The importance of legumes is evidenced by their high representation in *ex situ* germplasm collections, with more than 1,000,000 accessions worldwide. A detailed knowledge of the phylogenetic relationships of the Fabaceae is essential for understanding the origin and diversification of this economically and ecologically important family of angiosperms. This aims to combine the phylogenetic and genetic diversity approaches to better illustrate the origin, domestication history and preserved germplasm of major legume crops and to indicate further potential both for science and agriculture (Mohapatra *et al.*, 2023).

As with other crops like cereals, legumes face reduced genetic diversity, which impact negatively on the production of newly improved varieties showing stress tolerance. Conservation of wild genetic resources and cultivated germplasm will provide genetic materials for future breeding programmes. Genome sequencing libraries and bioinformatics tools could be used to screen and select genotypes with desirable traits, even according to the geographical patterns. The results will be of major importance for conservation genetics and breeding of newly improved cultivars that exhibit high resilience to adverse global climate patterns and plant pathogens.

The delineated molecular signatures regulating traits of agronomic importance have been utilized for introgression, combining, and pyramiding into selected grain legume crop genotypes of interest through traditional and advanced genomics-assisted breeding approaches in order to develop superior crop varieties in terms of yield and stress tolerance (Varshney, 2016). Recently, a chickpea cultivar 'Pusa JG16' has been released in India as a drought-hardy cultivar; this was developed through genomics-assisted breeding utilizing a QTL-hotspot region from ICC4958 (Bharadwaj *et al.*, 2022).

Varshney *et al.* (2021) have identified superior haplotypes through whole genome sequencing (WGS) of 3,336 accessions, both cultivated (3,171) and wild (195), for important traits relating to yield enhancement; these can then be introgressed into elite chickpea cultivars. The study also identified target genomic regions for the purging of deleterious alleles, which can be achieved through genomics-assisted breeding and/or gene editing (Varshney *et al.*, 2021). Similarly, superior haplotypes have been identified in pigeon pea based on the 292 pigeon pea genotypes of a reference set that included breeding lines, landraces, and wild species (Sinha *et al.*, 2020). These models are then used to predict the true breeding values of selecting particular candidates based on multi-omics data, big data technology, and artificial intelligence (Xu *et al.*, 2022). Thus, genomics plays an integral role in improving genetic gain in modern agricultural practices and has immense potential to expedite future grain legume crop improvement programs. The traditional and novel genomics-assisted breeding approaches that are now available provide clues to the quantitative dissection of complex trait regulation, and thus have potential to expedite studies of the genetic enhancement of complex traits in diverse grain legume crops.

Fabaceae (Leguminosae), with 800 genera and 20,000 species, is the third largest family of flowering plants, after the Orchidaceae and Asteraceae. It is an extremely diverse family with worldwide distribution, encompassing a broad range of life forms, from arctic alpine herbs and temperate or tropical perennial shrubs to annual xerophytes and equatorial giant trees. Reconstructing the phylogenetic relationship of the Fabaceae is essential to understanding the origin and diversification of this family. The conserved germplasm is characterized for distinct morpho-agronomic traits, using sets of crop-specific descriptors. Approximately 78% of the 146,837 grain legume germplasm accessions held in CGIAR centers have been characterized for morphological traits, including for resistance to biotic and abiotic stresses. However, only a small percentage of these collections have been characterized for biochemical traits. More emphasis and funding are needed in order to generate data on biochemical characteristics associated with biotic and abiotic stresses (Upadhyaya *et al.*, 2011). The Global Crop Diversity Trust (GCDT) supports the development of global crops (including legumes) and regional strategies for *ex situ* conservation and utilization of crop diversity. These strategies represent a major investment in the field of plant genetic resources (PGR), mobilizing experts to collaboratively plan for the more efficient and effective conservation and the use of crop diversity. The themes viewed under these strategies include regeneration, crop wild relatives (CWR), collecting, crop descriptors, information systems, user priorities, new technologies and research, and challenges to building a strategy for rational conservation. There is an urgency to ensure that the diversity in landraces is sampled and conserved in *ex situ* genebanks, especially as farming evolves from subsistence to a market-orientated endeavor dominated by modern cultivars and resulting in the erosion of crop genetic diversity (Sarmah *et al.*, 2024). A similar caution also applies to the *in situ* populations of wild relatives, as land use intensifies with urban expansion and climate change threatens ecology. A further development in targeting germplasm for key traits is to use a GPS map of landrace collection sites, which can be overlaid with climate data corresponding to vegetative and reproductive growth stages, and to identify landraces corresponding to sites with severe abiotic stresses during the previous 25 years. The landraces encompass much of the original diversity, plus accumulated mutations and genetic recombinations since domestication. These landraces are now disappearing, replaced by improved, higher yielding modern cultivars from modern breeding programs, as exemplified on recent collecting missions in China. On the premise that such landraces may have had natural selection for tolerances to these stresses, Li *et al.* (2013) identified pea landraces from China as priority candidates to evaluate for tolerance of reproductive frost stress, reproductive heat stress, and drought tolerance.

Adoption of the Germplasm Resource Information Network (GRIN-Global) database by international centers and national gene banks will allow for online queries across multiple gene banks for client-selected accessions, including multi-trait queries.

With a wide range of approaches now available for genotyping, and with the declining cost of whole genome sequencing, the greatest limitation for gene banks is precise phenotyping, not only for descriptive traits, but also for agriculturally relevant quantitative traits relating to the expression of yield, crop growth and disease resistance. To increase precision, a single seed should be used for self-pollination to provide genetically uniform progeny for genotypic and phenotypic analysis. This level of precision is desirable if the key alleles of genes for important agronomic traits are to be identified, but broad characterization of diversity in germplasm can be based on a pooled

DNA sample and phenotyping done on the bulked landrace mixture. Multi-environment analysis of quantitative variation involving multi-trait evaluation is far more informative than a single-environment trial and potentially provides some prediction for performance in other environments. The challenge for gene bank curators is to strategically sample collections and maximize information from costly evaluation trials. One approach is to use core collections, geographically sub-sampled or sampled using molecular marker diversity to characterize species diversity, or to sample based on priority traits. This has led to the use of climatic site descriptors for characterization of natural selection, focusing on abiotic stress response, and has therefore provided lists of prospective germplasm with potential tolerances to heat, frost, and drought stresses. With current advances in genotyping and phenotyping methods, it is possible to effectively mine and explore the diversity stored in germplasm collections (McCouch *et al.*, 2013).

Botanical Gardens and Herbariums as Sources of Information and Complementary Conservations

In addition to gene banks, botanical gardens offer an *ex situ* alternative to seed conservation. Botanical gardens face both challenges and opportunities in responding to global trends and, in particular, to climate change. The increased number of species at risk as a result of the changing climatic conditions will force many botanical gardens to refocus, to strengthen their conservation policies and to increase their participation in recovery programs for critically endangered species, including crop wild relatives (CWR). In addition, botanical gardens face an unprecedented opportunity to develop their role as introduction centers and play a major role in the assessment of new germplasm, both of ornamentals as well as other economically important plants (Heywood, 2011). Historically, botanical gardens have played a major role in plant introduction with far-reaching impacts and have been major drivers in human population growth and economic development of crops. Many botanical gardens manage seed banks of horticultural and wild species (such as the Millennium Seed Bank managed by Royal Botanic Garden at Kew, UK), have well-curated herbarium collections, are involved in re-introduction programs, and contain DNA storage facilities (DNA banks). Although herbarium and DNA banks are of relatively little practical use to conserve diversity, both provide valuable resources for studying CWR genetic diversity and information that can be used in gap analysis, as in the case of *Phaseolus* and *Lathyrus*.

In Situ Conservation of Crop Wild Relatives

In situ genetic reserve conservation may be defined as "the location, designation, management and monitoring of genetic diversity in natural wild populations within defined areas designated for active, long-term conservation". A genetic reserve is actively managed, even if the management involves only regular monitoring of the target CWR taxa; as long as the target population levels are maintained above the minimum viable population of approximately 5,000 individuals no further conservation action may be required. Importantly, *in situ* conservation action is a long-term commitment because significant resources have to be invested in order to establish a genetic reserve. Although the conservation goal is to always implement complementary conservation involving the parallel application of *in situ* and *ex situ* conservation techniques, there exists a preference for *in situ* conservation, primarily due to the overall need to maintain ecosystem health, but also because it has the advantage of maintaining the dynamic evolution of the CWR diversity itself in relation to parallel biotic and abiotic changes. Furthermore, due to the sheer number of CWR involved, the need to maintain effective genetic representation and the difficulty in precisely identifying which CWR or traits are required by plant breeders currently and in the future, *in situ* conservation is highly recommended, even if the main access route for breeders to diversity is via backup *in situ* samples deposited in *ex situ* gene banks.

All species in protected areas are passively conserved if the entire ecosystem or habitat is stable; however, without monitoring and active management, the genetic diversity within and between individual CWR populations could be eroded, and entire populations could even go extinct. It emphasizes that many protected areas already play an important role in the conservation of CWR species, even though many managers may be unaware that the land under their stewardship contains important crop genetic diversity. However, if our goal is to conserve the maximum genetic diversity within CWR taxa, then we need to study and monitor the genetic diversity and natural dynamics of CWR populations; otherwise, our efforts in establishing protected areas for these taxa may be wasted. It should also be noted that the *in situ* management of CWR may differ significantly from that required for more traditional protected areas whose objective is to sustain climax communities. For example, CWR of major crop plants are often located in pre-climax communities (*Lathyrus ervoides* Grande, *Lens orientalis* Popow, *Cicer bijugum* Rech. f.) where the site management is comparatively intense, or the CWR may be closely associated with traditional farming practices (*Vicia johannis* Tamamschjan, *Lathyrus cicera* L., *Pisumsativum* subsp. elatius Asch. & Graebn.), in which case, genetic reserve management would need to be associated with maintenance of the traditional farming/ranching system (Xu–Xiao *et al.*, 2008).Specifically, there has been very limited effort to conserve legume CWR diversity *in situ*. This has in part been due to two related disconnects: (a) the disconnect between academic studies identifying where genetic reserves or less formal *in situ*management activities should be established and their actual implementation, and (b) lack of collaboration between the plant genetic resource and protected area communities. Consequently, native legume populations are susceptible to genetic erosion or even extinction. Wild soybean (*Glycine soja* Wild.) is presumed to share a common ancestor with cultivated soybean.Apart from *ex situ* conservation, *in situ* strategy is also used to conserve wild soybean, since populations of *G. soja* typically show high levels of geneticheterogeneity. In China, more than 40 *in situ* conservation sites located in 15 provinces and regions have been established. Their genetic diversity is identified by genotyping 40 individuals at 20 SSR marker loci for each population, and the results showed that at least 90% of the total genetic diversity was present.

Common bean (*Phaseolus vulgaris* L.)

Several large germplasm collections for *Phaseolus* exist around the world, holding 261,968 accessions. The largest, with 36,000 accessions, is at the International Center for Tropical Agriculture (CIAT) based in Cali, Colombia. This collection includes a duplicate of the USDA collection, which consists of almost 18,000 accessions when considering all the species of *Phaseolus*, including wild and cultivated. Common beans make up the majority of the collection, and the representation of other species is limited in all of the collections. Other important *in situ* collections are found in Bolivia, Brazil and Mexico. Core collections have been made in both the CIAT and the USDA collections, while working collections are found in many other countries.

Cowpea (*Vigna unguiculata* (L.) Walp.)

Cultivated cowpea is divided into four cultivar groups: Biflora, Sesquipedalis, Textilis and Unguiculata. Cowpea belongs to culti-group unguiculata, while the yard-long bean or asparagus bean belongs to sesquipedalis. While cowpea is grown mainly for its dry grains in sub-Saharan Africa, South and Central America, the southern United States and Europe, the yard-long bean is commonly grown in Southeast Asia for the long, green, fleshy pods consumed as a vegetable. Because of its drought tolerance, cowpea is well adapted to the dry savanna; consequently, it is probably the most commonly grown and consumed legume in the dry savanna regions of sub-Saharan Africa. Most wild species of *Vigna* have adapted to various environments through the evolutionary process of diversification or specialization. They can provide an important gene pool for cultivated crops of *Vigna*.

Mungbean (*Vigna radiata* (L.) R. Wilczek)

Mungbean is a photo- and thermosensitive crop. The best temperature for its cultivation is 30–35°C with good atmospheric humidity. A total of 10,551 accessions of various *Vigna* species comprised of mungbean (3,704), urd bean (3,131), moth bean (1,486), rice bean (2,045) and azuki bean (185) have been stored at −18°C in the long-term repository of the national gene bank at NBPGR, New Delhi. Green gram germplasm accessions are maintained by more than 35 institutions globally, which hold a total of more than 25,000 accessions. IITA maintains over 15,000 accessions of cowpea, the Asian Vegetable Crops Research Centre, AVRDC at Taiwan, maintains 5,616 accessions of mungbean, and over 12,000 various *Vigna* are held in the Conservation Unit in Georgia. Limited germplasm accessions of moth bean are also available in several countries, including Bangladesh, Belgium and Kenya (Xu *et al*., 2008).

Genus *Glycine* Wild

The genus *Glycine* contains c. 22 species, which can be grouped in two subgenera. The first, subgenus Soja, includes two Asian annuals: *G. max* (the cultivated soybean) and the similar wild species *G. soja* Siebold & Zucc.

Soybean (*Glycine max* L.)

As the center of cultivated soybean (*G. max*), China has the most abundant genetic resources for soybean, with >23,000 cultivated soybean accessions and >7,000 wild soybean (*Glycine soja*) accessions conserved at the Chinese National Soybean GeneBank (CNSGB) and replicated at the National Germplasm Storage Facility in Qinhai.

In order to efficiently analyze and utilize this large *ex-situ* collection, a series of the core collections, including primary core, core, mini-core and integrated applied core collections-has been constructed based on the study of phenotypic and genotypic (SSR markers) datasets. In the past decades, these soybean core collections were widely used in the genomic study, molecular evolution clarification, elite genetic resources discovery, gene identification, elite lines development, and so on.

Genus *Cajanus* L.

The pigeonpea (*Cajanus cajan* L.) belongs to the legume subtribe Cajaninae. This subtribe contains a total of 13 genera. Until recently, the genera *Atylosia* Wight & Arn. and *Cajanus* were considered closely related genera, but van der Masen merged the genus *Atylosia* with *Cajanus*. In total, the combined genus *Cajanus* now has 32 species.

Pigeonpea (*Cajanus cajan* L.)

A total of 40,820 *Cajanus* accessions, comprising landraces, modern cultivars and CWR, have been conserved in *ex-situ* genebanks. The ICRISAT genebank holds 13,771 accessions, including 8,315 landraces, 4,830 breeding lines, 71 improved cultivars, and 555 accessions of wild relatives from 74 countries. Other genebanks conserving pigeonpea germplasm are the National Bureau of Plant Genetic Resources (11,427 accessions), New Delhi. As pigeonpea suffers limited genetic diversity within the cultivated gene pool, it is imperative to increase genetic diversity by using wild relatives from different gene pools. A number of species from the secondary gene pool (*C. sericeus*, *C. albicans* (Wight &Arn.) Maesen, *C. lineatus* (Wight &Arn.) Maesen, *C. trinervius* (DC.) Maesen, *C. cajanifolius* and *C. scarabaeoides*) have shown crossability with the cultivated type.

Lathyrus

In case of *Lathyrus*, among 1,555 accessions of 45 wild species conserved at ICARDA, a toxin-free gene has been identified in *L. tingitanus*, which can be used to develop toxin-free grass pea cultivars, providing its hybridization with *L. sativus* is possible. *Lathyrus* species, such as *L. ochrus* and *L. clymenum* and *L. cicero* have been identified as possessing resistance to *Orobanche*, a resistance that is not available within cultivated germplasm. *Lathyrus cicera* is also a good source for earliness and cold tolerance. However, alien gene transfer has hardly been attempted in grass pea in spite of the success of interspecific hybridization between *L. sativus* and two wild *Lathyrus* species (*L. cicera* and *L. amphicarpos* L.) with viable seeds. Other tested species formed pods but did not produce fully developed, viable seeds. It may be concluded that breeding strategies involving alien genetic transfer for the improvement of grasspea are possible through the readily crossable species *L. cicera* and *L. amphicarpos*, but biotechnology tools will be needed to assist in gene transfers among other species.

Wild Relatives as a Source of Novel Variation

Most researchers agree that wild relatives of crop plants are a useful source of novel variation for potential breeding (McCouch *et al.*, 2013). The challenges now are to reintroduce traits that have been lost or not used during the domestication process and subsequent breeding, including disease and resistance genes to make use of the wild allelic diversity that exists in germplasm collections. Highly variable germplasm is found in the secondary and tertiary pools of crop plants. This exotic material has largely remained uncharacterized and underutilized. Fortunately, there is a rising concern surrounding CWR use, and it is now a priority for GCDT. Genetic improvement of many crop plants has already benefited from the incorporation of traits from related wild species and other exotic germplasm sources. The development of pre-bred lines has long been advocated as a means of facilitating the transfer of genes from wild species. However, the majority of published results have been achieved with dedicated crosses and specific selection; thus they need to be made in trait-by-trait manner, which is a time-consuming and expensive process.

Conclusion

The improvement of grain legumes through genetic diversity is a dynamic and multidisciplinary process that involves traditional breeding methods, advanced biotechnological tools, and molecular techniques such as marker-assisted selection and genomic selection. By leveraging the vast genetic variability within and among legume species, breeders can create varieties that are better adapted to diverse environmental conditions and meet the growing demands for food, feed, and industrial uses. Thus, the conservation and utilization of genetic diversity in grain legumes are fundamental to ensuring their long-term sustainability and adaptability. A comprehensive approach that integrates conservation efforts, modern breeding techniques, and a better understanding of genetic resources will help develop improved legume varieties, thereby supporting global food security, agricultural sustainability, and the livelihoods of millions of people worldwide. Continued investment in research, collaboration among stakeholders, and policy support are crucial for fostering innovation and achieving sustainable improvement in grain legumes.

References

Bharadwaj, C., Jorben, J., Rao, A., Roorkiwal, M., Patil, B.S., Ahammed, S.K., et al. (2022). Development of high yielding Fusarium wilt resistant cultivar by pyramiding of "genes" through marker-assisted backcrossing in chickpea (Cicer arietinum L.). Front. Genet. 1747, 924287. doi:10.3389/fgene.2022.924287

Heywood, V.H. (2011). The role of botanic gardens as resource and introduction centres in the face of global change. Biodiv. Conserv. 20, 221–239.

Li, L., Redden, R.J., Zong, X., Berger, J.A. D. & Bennett, S.J. (2013). Ecogeographic analysis of pea collection sites from China to determine potential sites with abiotic stresses. Genet. Res. Crop Evol. 60, 1801–1815.

McCouch, S., Baute, G.J., Bradeen, J., Bramel, P., Bretting, P.K., Buckler, E., et al. (2013). Agriculture: Feeding the future. Nature 499, 23–24.

Mohapatra, P.P., Seleiman, M.F., Mandal, R., Pramanik, K., Maity, T.K., Tarafdar, J., et al. (2023). Efficiency of RAPD and SSR markers in assessing genetic diversity in summer onion (Allium cepa L.) genotypes. NotulaeBotanicaeHortiAgrobotanici Cluj-Napoca 51(3), 13369-13369.doi:10.15835/nbha51313369

Sarmah, D., Mohapatra, P.P., Seleiman, M.F., Mandal, T., Mandal, N., Pramanik, K., et al. (2024). Efficient regeneration of in vitro derived plants and genetic fidelity assessment of Phalaenopsis orchid. Frontiers in Sustainable Food Systems 8, 1359486. doi:10.3389/fsufs.2024.1359486

Sinha, P., Singh, V. K., Saxena, R. K., Khan, A. W., Abbai, R., Chitikineni, A., et al. (2020). Superior haplotypes for haplotype-based breeding for drought tolerance in pigeonpea (Cajanuscajan L.). Plant Biotechnol. J. 18(12), 2482–2490. doi:10.1111/pbi.13422

Upadhyaya, H.D., Dwivedi, S.L., Ambrose, M., Ellis, N., Berger, J., Smýkal, P.D., et al. (2011). Legume genetic resources: management, diversity assessment, and utilization in crop improvement. Euphytica 180, 27–47.

Varshney, R.K. (2016). Exciting journey of 10 years from genomes to fields and markets: Some success stories of genomics-assisted breeding in chickpea, pigeonpea and groundnut. Plant Sci. 242, 98–107. doi:10.1016/j.plantsci.2015.09.009

Varshney, R. K., Roorkiwal, M., Sun, S., Bajaj, P., Chitikineni, A., Thudi, M., et al. (2021). A chickpea genetic variation map based on the sequencing of 3, 366 genomes. Nature 599(7886), 622–627. doi:10.1038/s41586-021-04066-1

Xu, H., Jing, T., Tomooka, N., Kaga, A., Isemura, T. & Vaughan, D.A. (2008). Genetic diversity of the azuki bean (Vignaangularis (Willd.) OhwiandOhashi) gene pool as assessed by SSR markers. Genome 51(9), 728–738. doi:10.1139/G08-058

Xu–Xiao, Z., Jian–Ping, G., Wang, S., Liu, Q., Redden, R. R. & Ford, R. (2008). Genetic diversity and core collection of alien Pisumsativum L. germplasm. ActaAgron. Sin. 34(9), 1518–1528. doi:10.1016/s1875-2780(09)60003-1

Xu, Y., Zhang, X., Li, H., Zheng, H., Zhang, J., Olsen, M. S., et al. (2022). Smart breeding driven by big data, artificial intelligence and integrated genomic-enviromic prediction. Mol. Plant 15, 1664–1695. doi:10.1016/j.molp.2022.09.001

8

Grain Legumes as Climate Resilient Crops: Challenges and Prospects

Namitha Krishna*

Department of Agronomy, College of Agriculture, Vellayani, Kerala Agricultural University, Kerala, India

**Corresponding author's email: krishnanamitha09@gmail.com*

Abstract

Grain legumes, also known as pulses, are vital to sustainable agriculture due to their roles in enhancing soil fertility, providing essential nutrition, and supporting environmental sustainability. By establishing symbiotic relationships with nitrogen fixing bacteria, legumes naturally enrich soil with nitrogen, reducing reliance on synthetic fertilizers and fostering sustainable farming practices. Their inclusion in crop rotations not only boosts soil health but also supports food security, especially in developing countries where they are a key source of plant based protein and nutrients. The impact of climate change, including rising temperatures, altered precipitation patterns, and increased frequency of extreme weather events, presents significant challenges to agriculture. In this context, grain legumes demonstrate considerable climate resilience. Their drought tolerance, heat resistance, and ability to fix atmospheric nitrogen make them well-suited to withstand climate stress. Legumes exhibit mechanisms such as deep root systems, efficient water use, and adaptability to diverse environmental conditions, contributing to their robustness in the face of climate variability. To bolster the role of grain legumes in climate-resilient agriculture, climate-smart practices and technological advancements are essential. These include optimizing planting strategies, adopting water-efficient methods, and breeding for improved resilience traits. Addressing challenges such as climate variability, pest pressures, and resource limitations is crucial for maintaining the benefits of legumes. Future efforts should focus on developing robust management practices and improved varieties to ensure the continued contribution of grain legumes to food security and environmental sustainability.

Keywords: *Climate Resilience, Drought Tolerance, Grain Legumes, Heat Resistance, Sustainable Agriculture, Water Use Efficiency*

Introduction

Grain legumes, often referred to as pulses are integral to agriculture due to their multifaceted contributions to both the environment and human wellbeing. One of their most significant roles is in enhancing soil fertility. Through a symbiotic relationship with nitrogen fixing bacteria (rhizobium), legumes convert atmospheric nitrogen into usable form by plants, thereby naturally enriching the soil. This process reduces the reliance on chemical fertilizers, making farming practices more sustainable and cost effective. Furthermore, when included in crop rotation systems, legumes improve the yield and health of subsequent crops by replenishing essential nutrients in the soil. Besides their agricultural benefits, grain legumes are a cornerstone of human nutrition, particularly in developing countries. They are a rich source of plant-based protein, dietary fibre, essential vitamins, and minerals, offering a nutritious alternative where animal protein is scarce or too costly. This nutritional value makes them crucial for food security and combating malnutrition. Economically, grain legumes are a valuable crop for smallholder farmers. They not only provide a reliable source of food but also serve as an important cash crop, contributing to the livelihood of rural communities.

Legumes can promote environmental stability. Their ability to improve soil structure, prevent erosion, and enhance water retention contributes to better land management and resilience against environmental degradation. The term "carbon footprint" has gained significant attention in public discussions concerning responsibility and efforts to combat global climate change. Its usage has surged dramatically in recent months and years, becoming a popular word in the media, government, and business sectors. The carbon footprint is a measure of the exclusive total amount of carbon dioxide emissions that is directly and indirectly caused by an activity or is accumulated over the life stages of a product (Wiedmann and Minx, 2008). Moreover, compared to nitrogen intensive crops, legumes have a lower carbon foot print, making them a more environmentally friendly option in agriculture. Finally, in the context of climate change any legume species are resilient to harsh conditions such as drought and poor soils. This adaptability makes them vital for maintaining agricultural productivity in the face of changing climate patterns. They are also indispensable in promoting sustainable agriculture, improving human nutrition, supporting rural economies, and contributing to environmental stewardship.

Overview of Climate Change and Its Impact on Agriculture

Climate change refers to long term alterations in temperature, precipitation patterns and other atmospheric conditions primarily driven by human activities such as burning of fossil fuels, deforestation, and industrial processes. These activities increase the concentration of greenhouse gases in the atmosphere, leading to global warming. The impact of climate change on agriculture is significant and varies across regions. Rising temperatures can accelerate crop growth but may also reduce the growing period, leading to reduced yield. Many crops are sensitive to high temperatures, and heat stress, particularly during critical growth stages like flowering, that can severely reduce crop yields and quality. Changes in rainfall pattern is also a big concern. More frequent and intense droughts can lead to soil moisture deficits, making it difficult for crops to grow. Conversely, excessive rainfall and flooding can damage crops, erode soil, and disrupt planting and harvesting schedules. Additionally, climate change is expected to alter the distribution and behaviour of pests and diseases, potentially increasing crop losses. Extreme weather events, such as storms and heatwaves, pose further risks to agricultural productivity. Overall, the impact of climate change on agriculture is complex, with potential for both positive and negative outcomes, depending on how effectively adaptation strategies are implemented.

Importance of Climate Resilience in Modern Agriculture

Climate resilience in agriculture is essential for maintaining the sustainability and stability of food production systems in an era of increasingly unpredictable and extreme weather patterns. As climate change can disrupt global temperature, alters precipitation pattern, and increase the frequency of extreme weather events, agriculture must adapt to ensure continued productivity and food security (Ranjan *et al.*, 2023). Agrosystem resilience, refers to the ability of agricultural systems to adapt to social, economic, and environmental shifts by restructuring themselves. This resilience enables them to mitigate the effects of future changes, seize new opportunities, and ensure the ongoing sustainability of agrosystems (Folke *et al.*, 2010). The adoption of climate resilient practices helps farmers to make decisions and optimize resource use. Empowering farmers with knowledge about climate risks and adaptation strategies, supported by policy measures like subsidies and crop insurance, further strengthens resilience.

Gain Legumes as Climate-Resilient Crops

Grain legumes are a valuable gift from nature, and are rich in protein (16–50%), essential nutrients, dietary fibres, and vitamins. Grain legumes, such as lentils, chickpeas, beans, and peas are increasingly recognised as climate-resilient crops due to their ability to thrive in diverse and often challenging environmental conditions. These crops offer several advantages that make them particularly suited to withstand the impacts of climate change. Grain legumes occupy approximately 81 million hectares globally, yielding over 92 million tonnes annually. India contributes nearly a quarter of the world's annual grain legume production and is also the largest consumer. Other major grain legume-producing countries include China, Myanmar, Canada, Australia, Brazil, Argentina, the USA, and Russia (FAOSTAT, 2018). Grain legumes are naturally drought tolerant and can grow in semi-arid and arid regions where water is scarce. Their deep root systems enable them to access water from deeper soil layers, making them more resilient during periods of water stress. These legumes have unique ability to fix atmospheric nitrogen through symbiotic relationships with bacteria in their root nodules, reducing the need for synthetic nitrogen fertilizers. This not only lowers the carbon foot print of agricultural practices but also improves soil fertility, benefiting subsequent crops in rotation systems. Additionally, grain legumes are relatively resistant to temperature fluctuations and can adapt to a range of climatic conditions. Bayer *et al.* (2016) reported that legumes greatly enhance soil carbon retention under reduced tillage conditions, attributed to their higher crop biomass and the ability to moderate the rate of carbon mineralization compared to cereals. Grain legumes can transfer a significant portion of their photosynthetic carbon as root exudates including amino acids, phenolic acids, organic acids and various lignin rich compounds contributing largely to the C-sequestration and reducing C foot print (Hazra *et al.*, 2018). Alternatively, legumes enhance soil nitrogen levels for future crops and reduce global CO_2 emissions by over 300 Tg per year, potentially cutting emissions from nitrogen fertilizer (urea) production by up to 50 % (Jenser *et al.*, 2012). Legume rich feeds can significantly reduce methane (CH_4) emissions from the livestock industry. This is due to their lower fibre content, and the presence of condensed tannins and saponins, which decrease cell wall digestion and alter rumen methanogenesis (Eckerd *et al.*, 2010). Typically, plants with C_3 pathways thrive under elevated carbon dioxide conditions because they exhibit adaptive responses such as enhanced photosynthesis rates, a substantial increase in soluble sugars and starch, higher levels of growth promoting hormones like indole acetic acid, and improved symbiotic nitrogen fixation, which collectively make grain legumes superior to other plants (Palit *et al.*, 2020). Legumes can respond to climate change either positively or negatively, depending on their internal physiological adjustments and agricultural practices. The indeterminate growth nature of legumes complicates the task for climate scientists trying to determine the adaptation strategies or phenological triggers that support legumes in a changing climate. Addressing these diverse and complex factors requires experts to consider various stressors such as elevated CO_2, temperature changes, and biotic or abiotic challenges through agronomic, physiological, and microbiological research.

Mechanisms of Resilience in Grain Legumes

Drought Tolerance

Drought tolerance in grain legumes is crucial for ensuring crop productivity in water scarce regions. These plants exhibit several physiological adaptations, such as deep root systems that access water from deeper soil layers, efficient stomatal regulation to minimize water loss, and osmotic adjustment through the accumulation of osmolytes like proline, which helps maintain cell turgor during water stress. Throughout their evolution, plants, including legumes, have developed various drought resistance mechanisms to control their morphological and physiological traits, such as root and leaf structure, stomatal function, photosynthetic processes, and water accumulation. Plants that can thrive under moderate to severe drought conditions are considered "drought resistant" (Zhang *et al.*, 2022). To counteract the adverse effects of drought, legumes attract beneficial microbes that establish a symbiotic relationship with the host. Research indicates that the composition and diversity of these microbial communities differ across various plant parts, such as leaves, roots, and seeds. Enhancing drought tolerance in legumes can be achieved by managing these microbial communities through different strategies.

Legumes have developed a range of mechanisms to withstand drought stress, including modifications in their morphological and physiological traits, such as root and leaf architecture, stomatal control, and photosynthetic performance. These adaptations enable them to conserve water and sustain growth even under moderate to severe drought conditions. The microbiome plays a significant role in enhancing drought tolerance in legumes. Beneficial microbes can form symbiotic relationships with legume plants, helping them to better withstand drought conditions. The composition and diversity of the microbial community can vary across different plant parts, such as leaves, roots, and seeds, which can influence the plant's overall drought response. Legumes utilize a variety of adaptive mechanisms to mitigate the effects of drought. These include the synthesis of phytohormones and osmoprotectants, enhanced activity of beneficial microorganisms, and alterations in plant metabolism. Such mechanisms are crucial for maintaining cellular functions and overall plant health during periods of water scarcity (Petrushin *et al.*, 2023).

Heat Stress Resistance

Heat stress on grain legumes before flowering can lead to a decrease in germination rates, an increase in abnormal seedlings, early onset of flowering, and degeneration of nodules, which impairs nitrogen fixation. Additionally, it reduces membrane stability, photosynthetic and mitochondrial activities, and overall plant biomass. During the reproductive phase, heat stress negatively impacts pollen viability, fertilization, pod set, and seed development, often resulting in the abscission of flowers and pods and significant grain yield loss. Furthermore, heat stress can cause soil moisture deficits during reproductive growth stages, making grain legumes more susceptible to necrotrophic pathogens like *Rhizoctonia bataticola*, which causes dry root rot disease (Gaur, 2014).

Heat stress resistance in legumes is essential for sustaining crop yields in regions with elevated temperatures. These plants have developed various physiological and morphological adaptations to cope with heat stress. Physiologically, legumes produce heat shock proteins (HSPs) that protect cellular structures and ensure proper protein folding under stress. They also maintain cellular membrane stability to prevent damage and enhance the activity of antioxidant enzymes like superoxide dismutase (SOD) and catalase to combat oxidative stress. Morphologically, some legumes adjust leaf orientation to reduce heat absorption and water loss, while others have thicker leaves to minimize transpiration. Additionally, altering the timing of flowering to cooler parts of the day or season can help mitigate the effects of heat stress. Genetic approaches, such as selective breeding for heat tolerance and employing genomics and molecular breeding techniques, are also being utilized to develop cultivars that can maintain productivity under high-temperature conditions.

To effectively select for heat tolerance in grain legumes, crops can be planted in high-temperature areas or under late-sown conditions, followed by selecting plants or progenies based on the number of filled pods per plant and grain yield. Pollen-based screening can also be used to evaluate genotypes for heat stress tolerance. Genetic variation for heat tolerance exists in nearly all grain legumes, and it is important to utilize these diverse sources to develop heat-tolerant cultivars. The precision and efficiency of breeding programs can be improved by integrating advanced methods like marker-assisted selection, gametophytic selection, and precise phenotyping (Gaur, 2014).

Water Use Efficiency

Water use efficiency (WUE) is a crucial factor for enhancing climate resilience in legumes, particularly in the face of increasing droughts and water scarcity due to climate change. High WUE allows legumes to maximize biomass and yield production while minimizing water usage, making them more adaptable to fluctuating and adverse environmental conditions. Legumes with improved WUE are better equipped to maintain productivity during periods of limited water availability, thus contributing to food security in regions prone to drought. Additionally, high WUE in legumes can reduce the pressure on water resources, promoting sustainable agricultural practices (Kumar *et al.*,

2024). Breeding for higher WUE in legumes involves selecting traits that enhance water conservation, such as efficient root systems, optimized stomatal regulation, and sustained photosynthetic activity under water stress.

In semi-arid and arid tropical regions, where water availability is the most limiting factor for crop production, understanding and managing crop water requirements becomes crucial. Crops with lower water requirements are gaining importance as a key strategy for boosting food production in the face of water scarcity. Studies on water use in grain legumes have demonstrated that cowpea and faba bean have relatively low water consumption, with cowpea requiring between 78 and 258 mm of water and faba bean using between 101and 261 mm. Lentils can also be classified as low water users, particularly when compared to major grain legumes like dry beans, groundnut, and soybean, which have water requirements ranging from 318 to 463 mm, 697 to 809 mm, and 598 to 690 mm, respectively. The high water needs of groundnuts might also contribute to their increased susceptibility to aflatoxin contamination, as they are more prone to water-deficit stress. Therefore, it can be inferred that cowpea, faba bean, lentil, chickpea, and common pea are better suited for cultivation in arid and semi-arid regions where seasonal rainfall is low i.e., 200 to 400 mm (Chibarabada *et al.*, 2017).

Soil Fertility Improvement Through Nitrogen Fixation

Nitrogen fixing legumes play a crucial role in improving soil fertility through their symbiotic relationships with rhizobia, which transform atmospheric nitrogen into a plant-usable form. This process increases nitrogen availability and enhances soil organic carbon (SOC) levels. The cultivation of legumes stimulates microbial activity and improves soil structure through root exudates and the decomposition of organic residues. Legumes contribute around 50–70 Tg N ha^{-1} globally (Virk *et al.*, 2022), highlighting their vital role in sustainable agriculture by lessening the need for synthetic fertilizers. Additionally, legumes enhance soil health by boosting microbial diversity and enzymatic activities, which are crucial for nutrient cycling and maintaining soil quality (Sow *et al.*, 2023). The role of legumes in enhancing the soil fertility has been depicted in the Fig. 8.1.

Fig. 8.1: Benefits of legume crop in enhancing soil fertility (Stagnari *et al.*, 2017)

Legume plants and seeds are notably rich in protein, a characteristic directly linked to their ability to meet most of their nitrogen needs through symbiotic Rhizobia bacteria that live in their roots. When inoculated with the appropriate strain of Rhizobia bacteria, legumes can supply up to 90% of their own nitrogen (N). Shortly after a legume seed germinates in soil containing Rhizobia, the bacteria enter the root hairs and move into the root itself. The bacteria multiply, causing the root to swell and form pale pink nodules. These bacteria then bind nitrogen gas from the soil air while feeding on carbohydrates produced by the above-ground plant during photosynthesis. The bacteria convert nitrogen from the air and hydrogen from the plant's carbohydrates into ammonia (NH_3), which provides the plant with the nitrogen needed for growth. This symbiotic relationship between bacteria and legumes allows both to thrive and produce high-protein seeds or forage crops. Although legumes can fix nitrogen from the atmosphere, they can also absorb significant amounts of soil nitrogen if available. Nitrogen release from a legume crop occurs gradually as the above-ground plant residues, roots, and nodules decompose. Soil microorganisms break down the nitrogen-rich organic material and release nitrogen into the soil as they die. Typically, about two-thirds of the nitrogen fixed by a legume crop becomes available for the subsequent growing season when legumes are part of a crop rotation (Deakin and Broughton, 2009).

Resilient Grain Legumes

Cowpea: Cowpea contributes to soil fertility through nitrogen fixation, enriching the soil with nitrogen (70-350 kg ha^{-1}). This process not only benefits cowpea itself but also enhances the productivity of subsequent crops planted in the same soil (Dinah *et al.*, 2018). The genetic resources of cowpea are being actively explored to develop climate-resilient varieties. Advanced breeding techniques, including genomics-assisted breeding, are being employed to tap into the genetic diversity conserved at institutions like the International Institute of Tropical Agriculture (IITA). This research aims to identify traits that enhance drought and heat tolerance. As climate change leads to increased temperatures and variable rainfall patterns, cowpea's inherent traits make it a suitable candidate for adaptation strategies. Improved varieties of cowpea can be used as a major crop, while underutilized legumes like Bambara groundnut can serve as alternative crops to ensure food security.

African Yam Bean: African yam bean (AYB), known for its adaptability and resilience, is increasingly recognized as a valuable crop in the context of climate change. AYB can thrive in poor, marginal soils, including acidic and sandy types, which are often challenging for other crops. This adaptability makes it a suitable option for cultivation in regions where soil quality is compromised due to climate change (Giller, 2001). AYB exhibits a degree of drought and heat tolerance, which is crucial as climate change leads to increased temperatures and unpredictable rainfall patterns. This resilience allows AYB to maintain yields even under stress conditions, making it a reliable crop for food security in affected regions.

Bambara Groundnut: Bambara groundnut (BG) exhibits three primary drought tolerance mechanisms: avoidance, escape, and tolerance. It can adapt to water stress by reducing its vegetative growth period, flowering early, and maturing faster. This ability to respond to water scarcity is crucial in regions experiencing unpredictable rainfall and prolonged droughts. Research indicates that specific landraces of BG, such as 'Red' and 'Brown', show earlier maturity when subjected to water stress. For instance, these landraces mature at around 122.75 days after planting (DAP) under 30 % of crop water requirement, compared to 128 DAP under full water conditions. This adaptability allows BG to thrive even in challenging environments (Mabhaudhi *et al.*, 2013).

Winged Bean: Winged bean is known for its exceptional nutritional profile, with protein content comparable to that of soybean, ranging from 29.8 to 42.5 %. This high protein level is attributed to its effective nodulation and nitrogen-fixing capabilities, which enhance soil fertility and crop yield, making it a valuable food source in areas facing food insecurity due to climate change (Taofeek *et al.*, 2019). The winged bean thrives in heavy rainfall conditions, making it suitable for tropical equatorial countries in Africa. Its ability to grow in diverse environmental conditions allows it to adapt to the changing climate, particularly in regions experiencing increased rainfall variability and extreme weather events.

Climate Smart Agronomy for Grain Legumes

Climate smart agronomy for grain legumes involves applying agricultural practices designed to make legume production more resilient to climate change while reducing greenhouse gas emissions and enhancing productivity. This approach focuses on sustainably increasing yields by selecting climate-resilient legume varieties, optimizing planting dates, and using best agronomic practices like crop rotation and intercropping. It also emphasizes adapting to climate change through water management techniques such as drip irrigation and the use of drought-tolerant varieties, which help legumes withstand stressors like droughts and heatwaves. Additionally, practices like minimal tillage, biofertilizer use, and integrating legumes with agroforestry systems contribute to carbon sequestration and reduce the carbon footprint of legume farming. These strategies ensure food security under changing climatic conditions while promoting environmental sustainability.

Challenges

Grain legume production and cultivation face significant challenges that impact yield and sustainability. Climate variability and change, including unpredictable weather patterns like droughts and floods, can stress legume crops and reduce yields. Pests and diseases, such as aphids and fungal infections, pose ongoing threats, with new and resistant strains complicating management. Soil fertility issues, particularly in continuously cultivated lands, can lead to degradation despite legumes' benefits for soil health. Water scarcity during critical growth stages further limits production, particularly in regions with inadequate irrigation. Additionally, limited market access and poor infrastructure, including storage and transportation, can result in post-harvest losses and reduced profitability for farmers.

Future Prospects

a) **Development of climate smart management practices for grain legumes:** Climate smart management is essential for addressing the growing demand for food, fibre, and fuel in the face of climate change. Key practices include selecting drought-resistant and high-yielding crop varieties that are well-adapted to changing climates, implementing conservation agriculture, and adopting crop rotations and soil and water conservation technologies. Integrating tree-crop-livestock systems and using non-conventional energy sources like solar-powered drip irrigation also contribute to resource efficiency and resilience. Resource conservation technologies, along with integrated nutrient and pest management, are vital for adapting to emerging challenges Banerjee *et al.,* (2015) such as new insect pests. Establishing multi-stakeholder platforms is crucial for capacity building and raising awareness about these climate-smart practices. However, complexities in socio-cultural, political, economic, and ecological spheres can pose challenges to their adoption. Addressing these challenges require targeted interventions from both government and the private sector to promote climate-resilient management practices.

b) **Developing improved varieties suitable for inclusion in the new niches:** Genetically manipulating high yielding crop varieties is essential to combat emerging insect pests and withstand adverse climatic conditions. Preserving plant genetic resources and identifying locally available, genetically compatible hardy species for genetic manipulation is crucial to incorporating beneficial traits at the local level.

c) **Breeding for terminal heat and drought stress especially in winter grain legumes:** A thorough screening for terminal heat and drought tolerance across the extensive genetic material available in each grain legume is essential. Wild species and distant relatives of each legume should be examined for potential tolerance sources (Kumar *et al.*, 2019). Additionally, it is important to study the genetics of these traits within source germplasm or breeding populations. The use of molecular markers can aid in identifying and tracking specific genes or QTLs (quantitative trait locuses) of interest during the introgression process. Successfully incorporating heat or drought tolerance loci or QTLs into high-yielding grain legume genotypes will significantly enhance their adaptability across a wide range of cultivation areas.

d) **Breeding for biotic stress resistant varieties aggravated due to climate change:** Due to climate change, many previously minor diseases are becoming major threats, and certain insect pests are reaching epidemic levels that were once merely endemic or pandemic. It is crucial to consider various biotic stresses, such as disease and insect pest responses, when breeding for resilience to elevated CO_2 levels and rising temperatures. Genetic materials should be screened under artificial epiphytotic conditions or in the field with a high incidence of natural disease and pest pressure to ensure effective breeding against these climates related challenges.

Conclusion

In conclusion, grain legumes offer significant promise as climate-resilient crops due to their inherent adaptability and environmental benefits. Their role in enhancing soil fertility, providing essential nutrition, and supporting rural economies underscores their importance in sustainable agriculture. However, the impact of climate change presents considerable challenges, including increased biotic stresses, temperature extremes, and water scarcity. To address these issues, climate-smart management practices must be developed and adopted, focusing on selecting resilient varieties, optimizing agricultural practices, and incorporating advanced breeding techniques. By fostering multi-stakeholder collaboration, investing in genetic research, and improving resilience to both abiotic and biotic stresses, we can ensure that grain legumes continue to thrive and contribute to global food security in a changing climate. Addressing these challenges effectively will require coordinated efforts from governments, researchers, and farmers to enhance the sustainability and productivity of legume cultivation.

References

Adegboyega, T.T., Abberton, M. T., AbdelGadir, A. H., Dianda, M., Maziya-Dixon, B., Oyatomi, O.A., et al. (2019). Nutrient and antinutrient composition of winged bean (Psophocarpus tetragonolobus (L.) DC.) seeds and tubers. Journal of Food Quality, 1, 3075208.

Banerjee, H., Ray, K., Sarkar, S., Puste, A.M., Mozumder, M., Rana, L., 2015. Impact of nitrogen nutrition on productivity and nutrient use efficiency of potato (Solanum tuberosum) in an Inceptisol of West Bengal, India. SAARC J. Agric. 13(2), 141–150.

Banerjee, H., Sarkar, S., Ray, K., Rana, L., Chakraborty, A., 2016. Integrated nutrient management in potato-based cropping system in alluvial soil of West Bengal. Ann. Plant Soil Res. 18(1), 8–13.

Bayer, C., Gomes, J., Zanatta, J. A., Vieira, F. C. B., & Dieckow, J. (2016). Mitigating greenhouse gas emissions from a subtropical Ultisol by using long-term no-tillage in combination with legume cover crops. Soil and Tillage Research, 161, 86-94.

Chibarabada, T. P., Modi, A. T., & Mabhaudhi, T. (2017). Expounding the value of grain legumes in the semi-and arid tropics. Sustainability, 9(1), 60.

Deakin, W. J., & Broughton, W. J. (2009). Symbiotic use of pathogenic strategies: rhizobial protein secretion systems. Nature Reviews Microbiology, 7(4), 312-320.

Eckard, R. J., Grainger, C., & De Klein, C. A. M. (2010). Options for the abatement of methane and nitrous oxide from ruminant production: A review. Livestock science, 130(1-3), 47-56.

Folke, C., Carpenter, S. R., Walker, B., Scheffer, M., Chapin, T., & Rockström, J. (2010). Resilience thinking: integrating resilience, adaptability and transformability. Ecology and Society, 15(4), 43.

Gaur, P. M., Samineni, S., Krishnamurthy, L., Varshney, R. K., Kumar, S., Ghanem, M. E., & Babbar, A. (2014). High temperature tolerance in grain legumes. 6th International Food Legume Research Conference, Canada, Legume Perspectives, 7, 23499-23504.

Giller, K.E. (2001). Nitrogen Fixation in Tropical Cropping Systems. CAB Publishing, Wallingford, UK. 441.

Hazra, K. K., Ghosh, P. K., Venkatesh, M. S., Nath, C. P., Kumar, N., Singh, M., et al. (2018). Improving soil organic carbon pools through inclusion of summer mungbean in cereal-cereal cropping systems in Indo-Gangetic plain. Archives of Agronomy and Soil Science, 64(12), 1690-1704.

Jensen, E. S., Peoples, M. B., Boddey, R. M., Gresshoff, P. M., Hauggaard-Nielsen, H., JR Alves, B., & Morrison, M. J. (2012). Legumes for mitigation of climate change and the provision of feedstock for biofuels and biorefineries. A review. Agronomy for Sustainable Development, 32, 329-364.

Kirigia, D., Winkelmann, T., Kasili, R., & Mibus, H. (2018). Development stage, storage temperature and storage duration influence phytonutrient content in cowpea (Vigna unguiculata L. Walp.). Heliyon, 4(6), 24.

Kumar, J., Choudhary, A. K., Gupta, D. S., & Kumar, S. (2019). Towards exploitation of adaptive traits for climate-resilient smart pulses. International Journal of Molecular Sciences, 20(12), 2971.

Kumar, N., Sow, S., Rana, L., Kumar, V., Kumar, J., Pramanick, B., et al. (2024). Productivity, water use efficiency and soil properties of sugarcane as influenced by trash mulching and irrigation regimes under different planting systems in sandy loam soils. Frontiers in Sustainable Food Systems, 8, 1340551.

Mabhaudhi, T., & Modi, A. T. (2013). Growth, phenological and yield responses of a bambara groundnut (Vigna subterranea (L.) Verdc.) landrace to imposed water stress under field conditions. South African Journal of Plant and Soil, 30(2), 69-79.

Palit, P., Kudapa, H., Zougmore, R., Kholova, J., Whitbread, A., Sharma, M., & Varshney, R. K. (2020). An integrated research framework combining genomics, systems biology, physiology, modelling and breeding for legume improvement in response to elevated CO_2 under climate change scenario. Current Plant Biology, 22, 100149.

Petrushin, I. S., Vasilev, I. A., & Markova, Y. A. (2023). Drought Tolerance of legumes: Physiology and the role of the Microbiome. Current Issues in Molecular Biology, 45(8), 6311-6324.

Ranjan, S., Kumar, S., Dutta, S. K., Padhan, S. R., Dayal, P., Sow, S., et al. (2023). Influence of 36 years of integrated nutrient management on soil carbon sequestration, environmental footprint and agronomic productivity of wheat under rice-wheat cropping system. Frontiers in Environmental Science, 11, 1222909.

Sow, S., Singh, G., Ghosh, M., Dutta, S. K., Mandal, N., Kumar, S., & Ranjan, S. (2023). Nitrogen-management strategy through leaf-colour chart and SPAD meter for optimizing the productivity in irrigated wheat (Triticum aestivum). Indian Journal of Agronomy, 68(2), 219-222.

Stagnari, F., Maggio, A., Galieni, A., & Pisante, M. (2017). Multiple benefits of legumes for agriculture sustainability: an overview. Chemical and Biological Technologies in Agriculture, 4, 1-13.

Virk, A. L., Lin, B. J., Kan, Z. R., Qi, J. Y., Dang, Y. P., Lal, R., & Zhang, H. L. (2022). Simultaneous effects of legume cultivation on carbon and nitrogen accumulation in soil. Advances in Agronomy, 171, 75-110.

Zhang, H., Sun, X., & Dai, M. (2022). Improving crop drought resistance with plant growth regulators and rhizobacteria: Mechanisms, applications, and perspectives. Plant Communications, 3(1). 15p.

9

Grain Legumes for Nutritional Security and Health Benefits

Sudipta Sarker[1*], Hirak Banerjee[2] and Debjani Kundu[2]

[1]Department of Agronomy, Bidhan Chandra Krishi Viswavidyalaya, Mohanpur 741252, Nadia, West Bengal

[2]Regional Research Station (CSZ), BCKV, Akshaynagar, Kakdwip-743347, South 24 Parganas, West Bengal

**Corresponding author's email: sudiptasarker2015@gmail.com*

Abstract

Grain legumes, a group of crops which belong to Fabaceae family, are primarily grown for their edible seeds. Pulses contain a high amount of protein (20-25%), vitamins, and minerals. Consumption of pulses can be beneficial for cardiovascular health, weight management, for diabetics, cancer management, bone health etc. So, grain legumes can play a significant role to address future global food security, environmental sustainability, and nutritional needs. The nutritional value of grain legumes can be significantly enhanced through biofortification, agronomic intervention, processing, and value addition. So, diversification of food habit with pulses and cereals can solve the prolong problem of malnutrition in developing countries including India. Despite such great importance, global food legume production has been found to increase at a very slow pace.

Keywords: *Biofortification, Environmental sustainability, Food security, Grain legumes, Nutritional security*

Introduction

Grain legumes or pulses are a group of crops which belong to family Fabaceae, are primarily grown for their edible seeds. The common examples of grain legumes include: bean (*Phaseolus* spp.), lentil (*Lens culinaris*), chickpea (*Cicer arietinum*), pea (*Pisum sativum*), soybean (*Glycine max*), lupin (*Lupinus* spp.), broad bean (*Vicia faba*), igeon pea (*Cajanus cajan*), grass pea (*Lathyrus sativus*), black gram (*Vigna mungo*), green gram (*Vigna radiata*). Grain legumes can play a significant role to address future global food security, environmental sustainability, and nutritional needs. Pulses are the second most important food crop after cereals and an important source of proteins, vitamins, and minerals. Pulses are popularly known as "Poor man's meat" and "Rich man's vegetable". Pulses contain a high amount of protein (20-25%), which is almost double and triple of the protein content of wheat and rice, respectively and so can contribute significantly to the nutritional security of the country. Besides, they may serve the purpose of protein rich fodder and good quality green manure crop. Pulses can thrive well in low and marginal land with less external inputs. They act as erosion resisting crops because of its deep tap root system, leafy growth, and close spacing. Pulses can also be grown as *paira* crop or inter-crop in different growing situations. Being a leguminous crop, it can fix atmospheric nitrogen with the help of its root nodules, resulting in low requirement of external fertilizer. Keeping the importance of pulses in mind, United Nation's Food and Agricultural Organization (FAO) also declared 2016 as the 'International Year of Pulses'.

Being a rich source of protein, pulses along with cereals can provide one of the best alternatives to solve protein-calorie malnutrition problem in developing countries including India. Pulses contain 15 essential minerals which are important for human (Wang *et al.*, 2009). Consumption of pulses also decreases the chances of obesity and different metabolic syndrome (Rizkalla *et al.*, 2002). So, now-a-days several countries encourage people to consume pulses as part of a healthy diet.

Despite being so much important for human health and nutrition, global food legume production has increased marginally from 40.78 to 70 million tonnes (annual increase of 0.95 % only) except soybean and groundnut (Singh and Pratap, 2016). Besides slow growth of global grain legume production, other factors like increase in population, diversified uses of pulses, and enhanced purchasing capacity of people has put pressure on the per capita availability of grain legumes. The situation has further worsened due to lack of high yielding varieties (HYVs), low harvest index (HI), high susceptibility to diseases and insect pests, flower drops, lack of short duration varieties, intermediate growth habits, poor response to inputs and instabilities in performances of local varieties. To curved down these problems, few steps including proper pre- and postharvest management, efficient storage practice, value addition and manufacturing of quality nutritional products can be adopted.

Production Scenario of Grain Legumes

During 2022 the total world area under grain legumes cultivation was recorded as 959.68 lakh ha and total production was to the tune of 973.92 lakh tonnes with a productivity of 1015 kg/ha. Across the world, grain legumes are grown in almost 170 countries. Major pulse crops namely dry beans, chick pea, dry peas, pigeon pea and lentil have been found to grow in 105, 49, 95, 24 and 43 countries respectively, accounting 38, 15, 7, 6 and 6 of global pulse area, respectively. The highest share to world production was made by dry beans (29%) followed by chickpea (19%), dry peas (15%), lentil (7%), pigeon pea (5%) and others (15%) (Fig 1). India ranks 1st in chickpea, pigeon pea, dry beans production while 2nd in lentil and 4th in dry peas production.

India, Canada, China, Russian federation, Nigeria are the top 5 pulse producing countries in the world. In terms of area and production, India ranked first with a share of and 28% global area and production respectively. However, India's productivity is much lower (766 kg/ha) than the world average productivity (1015 kg/ha) (Table 1).

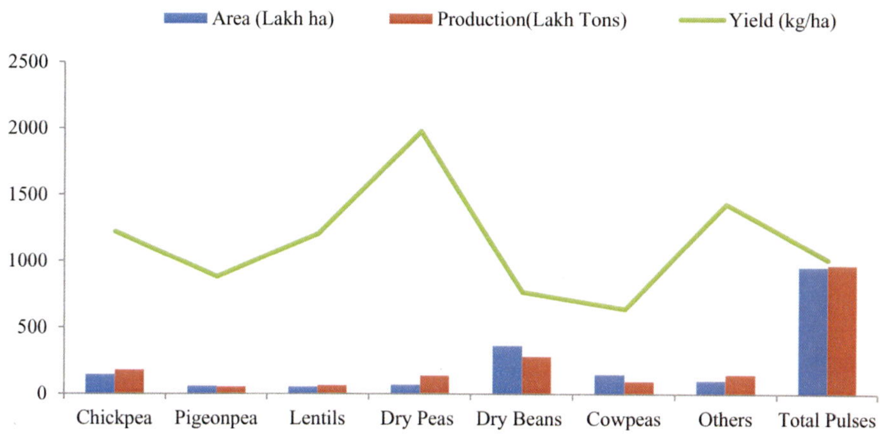

Fig. 9.1: Area, production and yield of different pulses in the world (Modified from: FAO, 2022)

Table 9.1: Country-wise pulse production scenario in the world during 2022 [Adopted from: FAO, 2022]

Sl. No.	Major countries	Area (Lakh ha)	Production (Lakh tonnes)	Productivity (kg/ha)
1.	India	361.11	276.69	766
2.	Canada	32.75	61.65	1882
3.	China	26.14	48.30	1848
4.	Russian Federation	23.08	46.07	1996
5.	Nigeria	49.25	42.01	853
6.	World	959.68	973.92	1015

Nutritional Composition

Various studies show, that there is large difference in different micro- and macronutrient content as well as in protein content among various pulses because of their genetic variability. Protein content in soybean, lentil, cowpea, chick pea, mungbean and pigeon pea varies from 26.5 to 57 % (Dwivedi *et al.* 1990; Burstin *et al.*, 2011).

Different pulses can provide energy from 93 to 432 kcal/100 g which is close to cereal grains. The energy mostly comes in the form of carbohydrates rather than fat. The carbohydrate content in grain legumes is about 60 % (Table 2). In grain legumes, mono- and oligosaccharides contribute a small percent of the total carbohydrate, while starch is the most prevalent carbohydrate. Generally, amylose content is more in grain legumes. Predominant sugars like stachyose, ajugose, raffinose, verbascose, and pentosans can be found in pulses. These sugars are not digested in the gut, come directly into the large bowel for fermentation which can cause abdominal discomfort and flatulence when pulses are eaten in large quantities. The content of these sugars can be reduced by overnight soaking of pulses. Pulses contain a low amount of fat (only 1%-6%) (Table 2). Soybean contains highest amount of fat (19%). Most of the fats are contributed by polyunsaturated and monounsaturated fatty acids. Pulses contain different kinds of minerals (1-4%) also (Table 9.2).

Table 9.2: Nutritional composition of grain legumes (per 100 g) [Modified from: TNAU Agritech]

Grain legumes	Energy (Kcals)	Moisture (g)	Protein (g)	Fat (g)	Mineral (g)	Carbohydrates (g)	Fibre (g)
Bengalgram	360-372	10-11	17-22	5-6	2-3	1-4	1-4
Blackgram (dal)	347	11	24	1	3	1	1
Cow pea	323	13	24	1	3	3	4
Field bean (dry)	347	10	25	1	3	1	1
Green gram (whole)	334-348	10	24	1	3	1-4	1-4
Horsegram (whole)	321	12	22	0	3	5	5
Grass pea (dal)	345	10	28	1	2	57	2
Lentil	343	12	25	1	2	59	1
Moth beans	330	11	24	1	3	56	4
Peas green	93-340	10-73	7-23	0-1	1-2	16-59	4
Rajmah	346	12	23	1	3	61	5
Redgram, dhal	116-335	13-65	10-22	1-2	1-3	17-58	1-6
Soyabean	432	8	43	19	4	21	4

The amount of nutrient significantly increases at the time of sprouting than the dried embryo. Sprouting increases nutritive value and the digestibility. Grain legumes also content good amount of dietary fibre (1-6%) especially when consumed with seed coat (Table 2). Pulse seeds contain higher dietary fiber (15-32 Total Dietary Fibre) than cereals among which around $1/3^{rd}$ to $3/4^{th}$ are insoluble and rest $2/3^{rd}$ to one quarter is soluble. Insoluble fiber promotes fecal bulking by water retention, while soluble fiber improves colon health by producing SCFA (Short-chain fatty acid), lowering pH, and slowing gastric empty rate.

Grain legumes are rich in vitamin B, specially thiamine, niacin, riboflavin, pyridoxine and pyridoxal (Table 3). Vitamin E, a strong antioxidant, can also be found in embryo of pulses. Pulse sprouts are also an important source of vitamin C. Interestingly, vitamin C content increases from negligible amount to 12 mg/100 g after 18 h of germination. Even tocopherol content is higher in pulses than cereals. Peas contain more α-than $\beta + \gamma$-tocopherols, and chickpea contains same amount of α- and $\beta + \gamma$-tocopherols.

Table 9.3: Vitamin composition in grain legumes (per 100 g) [Singh and Pratap, 2016]

Grain legumes	Thiamine (B_1) (mg)	Riboflavin (B_2) (mg)	Niacin (B_3)(mg)	Pantothenic acid(B_5) (mg)	Pyridoxine (B_6) (mg)	Folic Acid (B_9) (μg)	Vitamin C (mg)	Vitamin E (mg)
Chickpea	0.50	0.20	1.50	1.60	0.50	557	4.00	0.80
Pigeon pea	0.60	0.18	2.90	1.26	0.28	456	-	-
Lentil	0.80	0.20	2.60	2.21	0.54	479	4.40	0.30
Mung bean	0.60	0.20	2.20	1.90	0.40	625	4.80	0.50
Urdbean	0.60	0.20	2.30	-	0.20		4.80	
Field pea	0.70	0.20	2.90	1.80	0.20	274	0	0.30
Rajmash	0.53	0.22	2.08	0.79	0.40	399	1.80	-
Cowpea	0.94	0.22	2.36	1.39	0.44	545	4.60	-
Horsegram	0.40	0.20	1.50	-	-	-	-	-
Moth bean	0.40	0.09	1.50	-		-	-	-

The thiamine content is highest in cowpeas (0.94 mg/100 g), while lowest in horse gram and moth bean (0.4 mg/100 g). Rajmash and cowpea lead in riboflavin (0.22 mg/100 g). Pigeon pea and field pea are top in niacin content with a value of 2.9 mg/100 g (Table 3), whereas chickpea, horse gram, and moth beans each contain 1.5 mg/100 g. Lentils contain the highest pantothenic acid (2.21 mg/100 g), while rajmash has the lowest at 0.79 mg/100 g (Table 3). In terms of vitamin B6, lentil again is on top with a value of 0.54 mg/100 g, whereas, urd bean and field pea are at the lower end with a value of 0.2 mg/100 g. Mung bean is rich in folic acid (625 µg/100 g), while field pea has the least value (274 µg/100 g). Mung bean also lead in vitamin C with 4.8 mg/100 g, whereas rajmash has only 1.8 mg/100 g. The highest amount of vitamin E can be found in chickpeas (0.8 mg/100 g) (Table 3).

Pulses not only supply an adequate amount of vitamins but also supply significant amount of different minerals. Pulses are high in zinc (Zn), iron (Fe), magnesium (Mg) and phosphorus (P) and also contain more amount of calcium (Ca) than cereals. Pulses are also higher in potassium (K) which is beneficial for managing hypertension. They also contain some amount of manganese (Mn) and copper (Cu). Urd bean has the highest calcium content (154 mg/100 g), which is beneficial for bone health. Lentil has the lowest calcium content (69 mg/100 g.) Chickpea is rich in phosphorus (387 mg/100 g) which is essential for energy production and bone health, while lentil has the lowest phosphorus content (293 mg/100 g). The highest amount of iron presents in beans. Urd bean is particularly high in iron (9.1 mg/100 g) which is crucial for preventing anaemia. Mung bean is exceptionally high in magnesium (1771 mg/100 g) which is important for muscle and nerve function, while lentil contains only 94 mg/100 g (Table 4). Urd bean contains a high amount of zinc (16.2µg/g) and thereby supports immune function and wound healing. Pigeon pea has the lowest zinc content of 6.4 µg/g among pulses. Mung bean has the highest folic acid content (6.3 µg/g), which is crucial for DNA synthesis and repair. Except these minerals, lentil also contains Se (425–673 µg/kg) but it varies with location, growing conditions and soil characteristics. Only 100 g dry lentil can potentially provide 80–120 % of the minimum recommended daily Se intake. Selenium (Se) is a vital micronutrient in human nutrition which plays an important role in regulatory and protective functions.

Table 9.4: Mineral composition of major grain legumes [Soren *et al.*, 2016]

Grain legumes	Calcium(Ca) (mg/100 g)	Phosphorus (P) (mg/100 g)	Iron (Fe) (mg/100 g)	Magnesium (Mg) (mg/100g)	Zinc (Zn) (µg/g)
Chickpea	114	387	6.2	168	8.6
Pigeon pea	124	304	5.8	133	6.4
Mung bean	124	326	7.3	1771	11.7
Urd bean	154	385	9.1	185	16.2
Pea	75	298	5.1	124	-
Lentil	69	293	4.8	94	11.8

Chromium is another important mineral that participates in carbohydrate and lipid metabolism. The most common symptom of Cr deficiency is altered glucose tolerance which is related to diabetes and cardiovascular diseases, for which daily consumption of 0.05–0.2 mg Cr is recommended for adults. Generally, the level of Cr in pulses is about 0.05- 0.60 mg/g (Singh and Pratap, 2016).

Presence of Non-nutritive Bioactive Components in Pulses

Non-nutritive bioactive components are micronutrients found in plants that have biological activity and can have health benefits. They can also contribute to the flavour, colour, texture, and aroma of plants. Different types of phytochemicals are identified for their beneficial effect on human health, among which polyphenolic compounds, phytates, lectins and trypsin inhibitors are important. Isoflavones and lignans have antioxidant, weak oestrogenic and anticarcinogenic properties. Phenolic compounds including tannins can be found in pulse seed coat which has antioxidant activity. Phytoestrogens found in pulses can be effective t in preventing hormone-related cancers like breast cancer, prostate cancer etc. Phytic acid has DNA protective effect and antioxidant properties (Ranjan *et al.*, 2023).

In some pulses, lectins or hemagglutinins can be found which are toxic if taken orally. They can cause problems like vomiting, nausea, diarrhoea and bloating in humans. The lectins and enzyme inhibitors can decrease nutrient absorption rate and protein digestibility, respectively, but both have little impact after cooking. Phytic acid can reduce bioavailability of minerals. Some phenolic compounds also decrease protein digestibility and mineral bioavailability. Galactooligo saccharides may persuade flatulence. When lathyrus is consumed in large amounts the toxin (BOAA) present in lathyrus can cause lathyrism which is a neurological disorder.

Health Benefits: A Shield Against Diseases

The consumption of grain legumes has been linked to numerous health benefits, making them a vital component of a nutritious diet. Some key health benefits are discussed below:

- **Cardiovascular Health:** Regular consumption of grain legumes has been observed to lower cholesterol levels, reduce blood pressure, and improve overall heart health. The high fiber content helps in reducing LDL cholesterol by suppressing the synthesis of fatty acids by butyrate, acetate, and propionate (Crujeiras *et al.*, 2007), while the potassium content helps in managing blood pressure. Serum low-density lipoprotein (LDL) cholesterol and triglycerides are two major risk factors for coronary heart disease. Among the grain legumes, chickpea is the most hypocholesterolaemic agent. Epidemiological data indicates that consumption of pulse could lower the risk of coronary heart disease also (Chibbar *et al.*, 2010).
- **Weight Management:** Due to their high fibre and protein content, grain legumes can promote satiety and reduce overall calorie intake which aids in weight management. The protein, fibre, and slowly digested carbohydrate found in legumes may help to promote a feeling of fullness. According to National Health and Nutrition Examination Survey (NHANES) 1999-2002, Papanikolaou and Fulgoni (2008) found that adults who include a variety of grain legumes in their diet had significantly lower body weights compared with those who did not. Also grain legumes consumers were considerably less likely to be obese (BMI >30 kg/m^2) than non-consumers. They have a low glycemic index (GI) which helps in controlling appetite and preventing overeating.
- **Diabetes Management:** The complex carbohydrates and high fiber content in grain legumes contribute to better glycemic control, making them beneficial for persons with diabetes. They help in slowing down the assimilation of sugar, thereby preventing spikes in blood sugar levels. According to some studies, consumption of wholegrain foods protects against the development of type 2 diabetes mellitus. People who consume 3 or more servings of wholegrain foods on a daily basis are less likely to develop type 2 diabetes mellitus than the people who consume less (with a 20–30 % risk reduction). Pulses are typically low glycemic index (GI) foods (28 - 52) and hence, they usually have less impact on human blood glucose level On the hand, eating of low GI foods (<55) leads to moderate levels of glucose as compared to high GI foods (>70), which causes rapid increase in blood glucose (Atkinson *et al.*, 2008).

- **Digestive Health:** The insoluble fibre in grain legumes promotes regular bowel movements and prevents constipation. Additionally, the soluble fibre acts as a prebiotic which fosters the growth of beneficial gut bacteria and improves gut health.
- **Cancer Prevention:** Certain compounds found in grain legumes, such as flavonoids, lignans, and phytosterols, have been associated with a reduced risk of various cancers, including colorectal and breast cancer. Secondary metabolites which are derived from plants have anti-cancerous activities in different cancer types (Baskar *et al*., 2016). These compounds have antioxidant and anti-inflammatory properties that may help in inhibiting cancer cell growth. Pulses consumption regularly helps in lowering the risks of type 2 diabetes and also few cancers (Chibbar *et al*., 2010).
- **Bone Health:** The minerals in grain legumes, such as calcium (Ca), magnesium (Mg), and phosphorus (P), are essential for maintaining healthy bones. Regular consumption of legumes can help in preventing osteoporosis and other bone-related disorders.
- **Hypertension:** Legumes are rich source of potassium (K), magnesium (Mg), and fibre. All these nutrients have a positive impact on blood pressure management.
- **Celiac disease:** Celiac disease is a condition in which the absorptive surface of the small intestine is damaged by consumption gluten, which is commonly found in rye, wheat, and barley. Grain legumes possess higher amount of dietary minerals, proteins, vitamins, and fibres as compared to gluten-free cereal crops. Since pulses are rich in iron, and hence they can help celiac patients who often suffer from iron (Fe) and vitamin B deficiencies. Thus, pulse flours serve as a superior alternative to gluten-containing cereal flours which can supply valuable proteins and fibres in baked foods made from these flours.

Grain Legumes and Food Security

Grain legumes are a critical component of food security for several reasons as discussed below:

i) **Sustainability:** Grain legumes are often grown in rotation with other non-legume crops. Being a pulse crop, it can enhance soil fertility through symbiotic nitrogen fixation process in presence of *Rhizobium* bacteria mostly. This reduces the need for chemical fertilizers (particularly nitrogen fertilizers) and promotes sustainable agricultural development (Sow and Ranjan, 2021).

ii) **Climate Resilience:** Many grain legumes are drought-tolerant and can thrive in marginal soils, which make them adaptable to regions that are prone to climate variability. This resilience, as shown by most legume crops, helps in ensuring food production even under adverse climatic conditions. Being a short duration hardy crop with a deep tap root system, it can be grown successfully with low external inputs.

iii) **Affordability:** Grain legumes are generally affordable and accessible to low-income groups, which can provide an economical source of nutrition for them. Their long shelf-life also makes them a convenient food source, as these grains can be stored for extended periods without spoilage.

iv) **Dietary Diversification:** Incorporating grain legumes into daily-diets can help in diversifying food sources and reducing dependence on staple cereals. This diversification is essential for ensuring a balanced intake of nutrients on regular basis and preventing malnutrition in the long run.

Biofortification in Grain Legumes: A Possibility

In recent times, food security and malnutrition are the major two challenges experienced by developing countries including India. To achieve food security, global food production and productivity should be increased by introducing high yielding variety (HYVs) for most pulse crops. It is a need of an hour to adapt new technologies which can support sustainable production, ecological sustainability, and nutritional security. Pulse crops have high genetic diversity for micronutrients. They are highly adaptable to biofortification through traditional and marker-assisted breeding programmes, which will be highly beneficial in combating the malnutrition problem. Moreover, India is considered to be the biodiversity hotspot for pulses. This variability will serve as a future repository of gene/QTLs that control important traits for disease resistance, abiotic stresses, and nutritional quality. In India, the plant genetic stocks are being maintained by the National Bureau of Plant Genetic Resources (NBPGR), New Delhi. Out of total grain legume germplasms (69600), NBPGR maintains plant genetic resource of chickpea (17,449), pigeon pea (11,692), lentil (7,712), and mung bean (3,714) for subsequent use in biofortification process.

Recent Advances of Biofortification in Various Grain Legumes

- **Chickpea:** Chickpea is a rich source of iron (Fe), zinc (Zn), calcium (Ca), magnesium (Mg), copper (Cu), selenium (Se), and phosphorus (P). It has a wide genetic variation for most of the mineral micronutrients (Jukanti *et al.*, 2012). Iron content in seed may ranges from 3.0 to

14.3 mg and 2.2 to 20 mg of zinc in chickpea genotypes (Ray *et al.*, 2014). Diapari *et al.* (2014) found a market-trait association for Fe and Zn concentration where 94 different genotypes of chickpea were used including *'Desi'* and *'Kabuli'* type. According to Ray *et al.* (2014) the environment has a considerable impact on the mineral micronutrient content of field pea, common bean, chickpea and lentil cultivars.

- **Lentil:** Selenium (Se) is an important micronutrient which is a constituent of 25 human enzymes. Food and Nutrion Board (2000) recommended consumption of 55 µg of selenium/ day/ adult for healthy living. Lentil contains good amount of Se. An anti-nutrient compound, Phytic acid is present mainly on lentil seeds which has the capability to bind with mineral micronutrients and gradually can reduce their bioavailability. Thavarajah *et al.* (2009) reported lower phytic acid content genotypes of lentil (2.5–4.4 mg/g phytic acid).
- **Cowpea:** Cowpea couple lines were biofortified for high protein and mineral content, including potassium (K), calcium (Ca), iron (Fe), and zinc (Zn) (Singh, 2007).

Agronomic Fortification in Grain Legumes

Agronomic fortification is a strategy to increase the nutritional quality of crops by increasing the concentration of essential micronutrients through agricultural practices. This approach is particularly important for pulses, which are a staple food in many regions and a key source of protein and other nutrients. Application of micronutrient-rich fertilizers to the soil or as foliar sprays can enhance the nutrient content in field crops. Improving soil health through application of organic amendments and proper irrigation practices can also enhance nutrient uptake by plants

Iron and Zinc Fortification in Chickpeas: According to Shivay *et al.* (2016), soil application of iron (Fe) and zinc (Zn)-enriched fertilizers enhanced iron (Fe) and zinc (Zn) concentrations in chickpea grains and improves their overall nutritional quality.

Zinc Fortification in Pigeon Peas: Foliar application of zinc sulfate solution increased zinc content in pigeon pea seeds which contributes in better dietary zinc intake (Hazra and Bohra, 2021).

Iron and Zinc Fortification in Lentil: According to Dhaliwal *et al.* (2021) combined application of $ZnSO_4$, $7H_2O$ (0.5%) and $FeSO_4$, $7H_2O$ (0.5%) at pre-flowering stage can significantly increase yield, uptake of Fe and Zn, and also gives good economic returns.

Government of India Initiatives

During last two decades the Government of India (GoI) has implemented several pulse-related schemes and taken initiatives to promote the production, consumption, and nutritional benefits of grain legumes. These schemes aim to enhance food security, improve farmers' livelihoods, and address nutritional deficiencies. Some key government schemes and programs which are related to grain legumes are given below:

i) **National Food Security Mission (NFSM):** The scheme 'National Food Security Mission' (NFSM) was launched in 2007-08 with an objective to increase the production of rice, wheat, and pulses. The mission focuses on expanding the area under cultivation, enhancing productivity, and ensuring efficient use of resources. Under this scheme, a specific component was NFSM-Pulses which deal with pulse crop only. Under this component, the mission promotes the cultivation of pulses through various interventions such as the distribution of high-yielding varieties, improved agronomic practices, and pest management strategies. Financial assistance is also given to farmers for adopting these practices.

ii) **Integrated Scheme on Oilseeds, Pulses, Palm oil, and Maize (ISOPOM):** The scheme 'ISOPOM' was launched in 2004-05 with an objective to increase the production as well as productivity of oilseeds, pulses, palm oil, and maize. This scheme provides support for seed distribution, pest control measures, and farmer training programs. The pulses development component under ISOPOM focuses on providing quality seeds, promoting integrated nutrient and pest management, and conducting farmers' field demonstrations to showcase improved practices.

iii) **Pradhan Mantri Fasal Bima Yojana (PMFBY):** The scheme was launched in 2016. The 'PMFBY' is a crop insurance scheme which aims at providing financial support to farmers in the event of crop failure due to different causes like natural calamities, pests- diseases. The scheme not only covers pulses but also a wide range of crops..The specific component 'Crop Insurance for Pulses' under the insurance scheme, PMFBY helps protect farmers from potential financial losses and encourages them to invest in pulse cultivation.

iv) **Paramparagat Krishi Vikas Yojana (PKVY):** The 'PKVY' scheme promotes organic farming in India through a cluster-based approach. The scheme aims to improve soil health, reduce input costs, and enhance the earning of farmers. The specific componet'Organic Pulses' Under

PKVY, encourages farmers to grow organic pulses, which can fetch higher market prices and contribute to sustainable agricultural practices.

v) **Rashtriya Krishi Vikas Yojana (RKVY):** The 'RKVY' scheme is a comprehensive scheme which aims at achieving 4% annual growth in agriculture through the development of agriculture and allied sectors. The scheme gives financial assistance to states for implementing various agricultural development projects, including those related to pulses. Under 'Pulse Projects', States can propose and implement projects focused on enhancing pulse production, improving post-harvest management, and developing market linkages under RKVY.

vi) **Mid-Day Meal Scheme (MDM):** The Mid-Day Meal Scheme aims to improve the nutritional status of school children by providing them with a balanced meal during school hours. Pulses are a key component of the meals provided under this scheme. By including pulses in the mid-day meals, the scheme ensures that children receive adequate protein and other essential nutrients, contributing to their overall growth and development.

vii) **Public Distribution System (PDS):** The Public Distribution System (PDS) is a government-managed food distribution network aimed at providing subsidized food grains to the poor. Pulses are often included in the list of subsidized commodities distributed through PDS. By making pulses available at subsidized rates, PDS helps improve the nutritional intake of low-income households and addresses protein deficiencies.

viii) **National Agricultural Market (e-NAM):** The National Agricultural Market (e-NAM) is an online trading platform which aims at creating a unified national market for agricultural commodities. It facilitates better price discovery, transparency, and ease of doing business for farmers. Through e-NAM, pulse farmers can access a wider market, compare prices across different markets, and sell their produce at competitive rates.

ix) **Soil Health Card Scheme:** The scheme was launched in 2015, which aims to provide farmers with information on the nutrient status of their soil and recommendations for appropriate soil management practices. By promoting balanced nutrient management, the scheme helps improve the productivity of pulse crops and ensures sustainable soil health.

x) **Pradhan Mantri Kisan Samman Nidhi (PM-KISAN):** The 'PM-KISAN', launched in February 2019, is a direct income support scheme that provides financial assistance to small and marginal farmers. Under this scheme, eligible farmers receive annual financial support to meet their agricultural and household needs. The financial assistance provided under PM-KISAN can help pulse farmers invest in quality seeds, fertilizers, and other inputs, thereby improving their productivity and income.

Conclusion

Food and nutritional insecurity is highly prevalent in developing and resource-constrained nations. Diversifying food systems and diets not only enhances health but also contributes to numerous additional benefits that supports healthy ecosystems. Grain legumes stand out as a robust source of vegetable proteins and minerals, making them a good alternative for diversifying a cereal-based diet. Pulses with cereals can offer a way to solve the protein calorie malnutrition problem in the developing nations including India. So, this diversification is particularly crucial for ensuring balanced food and nutrition security in poorer regions. If given proper attention, it has the potential to eradicate this prolonged problem of malnutrition from our world.

Moreover, the nutritional value of grain legumes can be significantly enhanced through biofortification, processing, and value addition. Dietary modification and nutrient enhancement, along with the promotion of their sustainable consumption, further increase their benefits. By implementing these strategies, the role of grain legumes in food security and nutritional improvement can be maximized, ultimately contributing to the well-being of communities in developing nations.

References

Atkinson, F.S., Foster-Powell, K., & Brand-Miller, J.C. (2008). International tables of glycemic index and glycemic load values: 2008. Diabetes Care, 31(12), 2281-2283. doi:10.2337/dc08-1239.

Baskar, V., Park, S.W., & Nile, S.H. (2016). An update on potential perspectives of glucosinolates on protection against microbial pathogens and endocrine dysfunctions in humans. Critical Reviews in Food Science and Nutrition, 56(13), 2231–2249. doi:10.1080/10408398.2014.910748.

Burstin, J., Gallardo, K., Mir, R.R., Varshney, R.K., & Duc, G. (2011). Improving protein content and nutritional quality. In Biology and breeding of food legumes, Pratap, A. & Kumar. J. (Eds), CABI, pp. 314–328.

Chibbar, R.N., Ambigaipalan, P., & Hoover, R. (2010). Molecular diversity in pulse seed starch and complex carbohydrates and its role in human nutrition and health. Cereal Chemistry, 87(4), 342–352.

Crujeiras, A.B., Parra, D., Abete, I., & Martinez, J.A. (2007). A hypocaloric diet enriched in legumes specifically mitigates lipid peroxidation in obese subjects. Free Radical Research, 41(4), 498-506.

Dhaliwal, S.S., Sharma, V., Shukla, A.K., Kaur, J., Verma, V., Singh, P., Singh, H., Abdel-Hafez, S.H., Sayed, S., Gaber, A., Ali, R., & Hossain, A. (2021). Enrichment of Zinc and Iron Micronutrients in Lentil (Lens culinaris Medik.) through Biofortification. Molecules, 26(24), 7671.

Diapari, M., Sindhu, A., Bett, K., Deokar, A., Warkentin, T.D., & Tar'an, B. (2014). Genetic diversity and association mapping of iron and zinc concentrations in chickpea (Cicer arietinum L.). Genome, 57(8), 459–468. doi:10.1007/978-81-322-2716-8_10.

Dwivedi, S.L., Jambunathan, R., Nigam, S.N., Raghunath, K., Shankar, K.R., & Nagabhushanam, G.V.S. (1990). Relationship of seed mass to oil and protein contents in peanut (Arachis hypogaea L.). Peanut Science, 17(1), 48–52.

Hazra, K.K. & Bohra, A. (2021). Increasing relevance of pulse crops to sustainable intensification of Indian agriculture. National Academy of Sciences Letters, India, 44, 1-3. doi:10.1007/s40009-020-00948-6.

Jukanti, A.K., Gaur, P.M., Gowda, C.L.L., & Chibbar, R. N. (2012). Nutritional quality and health benefits of chickpea (Cicer arietinum L.): A review. British Journal of Nutrition, 108(S1), S11–S26.

Papanikolaou, Y., & Fulgoni, V. L., 3rd. (2008). Bean consumption is associated with greater nutrient intake, reduced systolic blood pressure, lower body weight, and a smaller waist circumference in adults: Results from the National Health and Nutrition Examination Survey 1999–2002. Journal of the American College of Nutrition, 27(5), 569–576.

Ranjan, S., Sow, S., Ghosh, M., Padhan, S. R., Kumar, S., Gitari, H., et al. (2023). Nutraceutical properties and secondary metabolites of quinoa (Chenopodium quinoa Willd.): a review. International Journal of Food Properties, 26(2), 3477-3491. doi:10.1080/10942912.2023.2286895.

Ray, H., Bett, K., Tar'an, B., Vandenberg, A., Thavarajah, D., & Warkentin, T.D. (2014). Mineral micronutrient content of cultivars of field pea, chickpea, common bean, and lentil grown in Saskatchewan, Canada. Crop Science, 54(4), 1698–1708.

Rizkalla, S.W., Bellisle, F., & Slama, G. (2002). Health benefits of low glycaemic index foods, such as pulses, in diabetic patients and healthy individuals. British Journal of Nutrition, 88(S3), 255–262.

Shivay, Y.S., Singh, U., Prasad, R., & Kaur, R. (2016). Agronomic interventions for micronutrient biofortification of pulses. Indian Journal of Agronomy, 61(Special Issue), 161-172.

Singh, B.B. (2007). Recent progress in cowpea genetics and breeding. Acta Horticulturae, 752, 69–75.

Singh, N.P., & Pratap, A. (2016). Food Legumes for Nutritional Security and Health Benefits. In Biofortification of Food Crops, Springer, pp. 41-50. doi:10.1007/978-81-322-2716-8_4.

Soren, K.R., Shanmugavadivel, P.S., Gangwar, P., Singh, P., Das, A., & Singh, N.P. (2016). Genomics-enabled breeding for enhancing micronutrients in crops. In Nutritional quality improvement in plants. Springer. pp. 151–170.

Sow, S., & Ranjan, S. (2021). A Review on agronomic biofortification for improving food and nutritional security. Research Biotica, 3(2), 139-44. doi:10.54083/ResBio/3.2.2021.139-144.

Thavarajah, P., Thavarajah, D., & Vandenberg, A. (2009). Low phytic acid lentils (Lens culinaris L.): A potential solution for increased micronutrient bioavailability. Journal of Agricultural and Food Chemistry, 57(19), 9044–9049.

TNAU Agritech. Tamil Nadu Agricultural University. Composition of pulses. Retrieved from https://agritech.tnau.ac.in/postharvest/pht_pulses_composition.html (Accessed on 1st August 2024).

Wang, N., Hatcher, D.W., Toews, R., & Gawalko, E.J. (2009) Influence of cooking and dehulling on nutritional composition of several varieties of lentils (Lens culinaris). Food Science Technology, 42, 842–848.

10

Global Perspective of Grain Legume Production, Trade and Policy Imperatives

Dhruvendra Singh Sachan*, M.Z. Siddiqui and Shivendra Singh

Department of Agronomy, College of Agriculture, Chandra Shekhar Azad University of Agriculture and Technology, Kanpur, U.P., India- 208 002

**Corresponding author's email: me.dhruv143@gmail.com*

Abstract

The rapid increase in global food demand, driven by fast population growth, has made it challenging to provide nutritionally balanced diets. Pulses offer a promising solution to these challenges by delivering significant nutritional and physiological benefits. Pulses such as chickpeas, green gram, peas, horse gram, beans, lentils, black gram, and others are excellent sources of protein (190-260 g kg^{-1}), carbohydrates (600-630 g kg^{-1}), dietary fibers, and bioactive compounds. The phytochemicals found in pulses, including flavonoids, phenolics, tannins, phytates, saponins, oxalates, peptides, and enzyme inhibitors, offer numerous health benefits. Pulses, essential to human diets for centuries, have seen historically low consumption levels due to underappreciation of their nutritional value. Recognizing their significance, the United Nations designated 10 February as World Pulses Day, and declared the year, 2016 as International Year of Pulses emphasizing their role in promoting healthy diets, sustainable food production, and food security. India is the largest producer of pulses in the world accounting for 25% of the global pulse production. The production and productivity of pulses have slightly decreased due to great emphasis to rice and wheat crops. To boost pulse production, India has implemented initiatives like the National Food Security Mission and policies to support farmers, including Minimum Support Prices (MSP) and welfare schemes integrating pulses. These efforts aim to enhance pulse cultivation, ensure food security, and improve farmer incomes, addressing the demand-supply gap and declining per capita availability of pulses.

Keywords: *legumes, sustainable, oxalates, inhibitors, phytochemicals*

Introduction

Pulses have been a vital component of the human diet for centuries, yet their nutritional benefits are often overlooked, leading to low average consumption. The United Nations General Assembly designated 10 February as World Pulses Day. The International Year of Pulses, 2016, played a significant role in changing these perceptions, highlighting the essential roles pulses play in promoting healthy diets, sustainable food production, and, most importantly, food security. As nutritious seeds for a sustainable future, pulses can significantly contribute to achieving many of the Sustainable Development Goals outlined in the 2030 Agenda for Sustainable Development.

Pulses are consumed in various forms, including whole seeds, split grains, dehulled split grains, flour, and sprouted seeds. They are the edible dry seeds from plants in the *Leguminosae* family. Although the terms "pulses" and "legumes" are often used interchangeably, there is a subtle distinction: while all pulses are legumes, not all legumes qualify as pulses (Singh, 2017). Pulses are also called as "mini nitrogen factories" (Singh, 2018). Pulses play a critical role in agriculture due to their numerous benefits for nutrition, soil health, economic stability, biodiversity, and environmental sustainability (Lovegrove *et al.*, 2023). Nutritionally, pulses like lentils, chickpeas, and beans are rich in proteins, fiber, vitamins, and minerals, making them essential in diets, particularly in developing countries where they are a primary source of protein (Singh, 2017; Venkidasamy *et al.*, 2019). From an agricultural perspective, pulses significantly enhance soil fertility through biological nitrogen fixation (Zahran, 1999). They form a symbiotic relationship with nitrogen-fixing bacteria, which converts atmospheric nitrogen into a form usable by plants, thereby reducing the need for chemical fertilizers and promoting sustainable farming practices. The symbiotic relationship between Rhizobium species and leguminous plants facilitates the fixation of atmospheric nitrogen (N_2), providing a renewable and sustainable source of nitrogen for agricultural systems (Peoples *et al.*, 1995a).

Estimates for nitrogen (N) fixation by various legume crops are often quite substantial, typically ranging between 200 and 300 kg of N per hectare per year (Peoples *et al.*, 1995b). The increase in main crop yields (legume and non-legumes) is up to 20% due to legumes (Zhao *et al.*, 2022). For crops like alfalfa, red clover, peas, soybeans, cowpeas, and vetch, the inputs of fixed nitrogen have been estimated to range from 23 to 335 kg of N per hectare annually (Wani *et al.*, 1995). Biological nitrogen fixation contributes around 122 million tonnes of nitrogen annually, with agricultural crops responsible for fixing approximately 58 to 62 million tons of this total (Rao and Balachandar,

2017; Vitousek *et al.,* 2013). Soybean (*Glycine max*) is the leading legume in BNF, fixing about 16.4 million tonnes of N annually (Hungria *et al.,* 2015) and the main bacteria involved in this fixation belongs to the *Bradyrhizobium* species (Gyogluu *et al.,* 2018).

Pulses are also valuable for their climate resilience (Benitez-Alfonso *et al.,* 2023). They can thrive in various environments, including arid and semi-arid regions, making them suitable for cultivation in areas with water scarcity (Nyaombo, 2022). This resilience contributes to agricultural sustainability in the face of climate change (Ranjan *et al.*, 2022). Economically, pulses offer advantages to smallholder farmers by providing higher market returns compared to cereals (Abraham and Pingali, 2021). Their ability to be intercropped with other crops maximizes land use, providing an additional income stream for farmers (Singh, 2018).

Fig. 10.1: Different types of grain legumes

Global Legumes Production Scenario

In the early 1970s, global pulse production made up approximately 3.5% of the total global cereal production. However, due to a slow growth rate over the past five decades, this ratio decreased to about 2.8 percent by the three-year period ending in 2014. During the early 1970s, the average yield per hectare for pulses was roughly one-third of that for cereals. As cereal crop yields grew at a faster pace, the gap between cereal and pulse yields expanded over the decades. By the triennium ending in 2014, the average yield for pulse crops had fallen to about a quarter of the average yield for cereal crops.

Globally, pulses are cultivated on approximately 94.14 million hectares, yielding around 89.74 million tonnes at an average productivity of 953 kg/ha. India leads the world in pulse production, with over 30 million hectares dedicated to their cultivation, making it the largest producer of pulses globally. India is the largest producer (25% of global production), consumer (27% of world consumption) and importer (14%) of pulses in the world.

Table 10.1: Global Pulses Production and projections

Global Pulses Production and Projections				
	Production ('000 tonnes)		**Growth (%)**	
	Average 2020-22 estimate	2032	2013-22	2023-32
World	91 540	120 551	1.71	2.34
North America	10 432	14 519	0.95	2.13
Canada	7 819	11 053	1.24	2.00
United States	2 613	3 466	0.13	2.58
Latin America	7 212	7 944	-0.25	1.18
Argentina	904	1 093	7.43	1.61
Brazil	2 853	3 010	-1.65	0.72
Chile	70	59	5.78	-0.74
Colombia	228	268	0.69	2.48
Mexico	1 320	1 577	-2.25	1.80
Paraguay	93	107	2.61	1.78
Peru	273	297	0.44	1.38
Europe	10 209	13 452	4.06	2.53
European Union	4 344	6 461	4.00	3.89
United Kingdom	995	1 145	5.11	1.10
Russia	3 549	4 432	4.98	1.74
Ukraine	773	798	5.98	0.49
Africa	22 144	28 589	1.70	2.62
Egypt	281	268	0.23	0.22
Ethiopia	2 960	3 291	-0.40	1.89
Nigeria	3 870	4 764	1.03	2.21
South Africa	88	111	0.22	2.32
Asia	39 001	53 142	2.03	2.43
China	5 333	5 918	2.68	0.75
India	22 837	33 884	2.90	3.03
Indonesia	155	155	-7.17	0.24
Iran	408	366	-6.67	-0.68
Japan	122	122	-0.49	0.23

Kazakhstan	643	924	28.18	3.80
Korea	15	23	-3.95	4.92
Pakistan	775	837	-0.53	0.43
Philippines	57	59	-2.70	0.81
Saudi Arabia	18	18	2.96	1.01
Thailand	209	238	-0.63	0.94
Türkiye	1 375	1 538	2.16	1.26
Viet Nam	339	393	0.87	1.43
Oceania	2 542	2 905	-0.55	1.42
Australia	2 508	2 869	-0.48	1.43
New Zealand	20	21	-5.01	0.75

Source: OECD/FAO (2023)

Global Ranking: Crop Wise Among Pulses

In 2021, the global area under pulse cultivation was approximately 954.39 lakh hectares, yielding around 889.67 lakh tonnes with an average productivity of 932 kg/ha.

Table 10.2: Global Ranking among Pulses (FAO statistics, 2021)

Crop	Area (lakh ha)	Area contribution(%)	Production (lakh tonnes)	Production contribution (%)	Yield (kg/ha)	India's Rank
Gram	150.05	16	158.72	18	1058	1st
Tur/Arhar	63.57	7	54.77	6	862	1st
Lentil	55.86	6	56.1	6	1004	2nd
Peas	70.44	7	124.04	14	1761	4th
Beans dry	359.21	38	277.15	31	772	1st
Cowpea	149.11	16	89.86	10	603	
others	106.16	11	129.03	15	1215	
Total Pulses	954.39		889.67		932	1st

Legumes Production Scenario of India

India is the world's leading producer, consumer, and importer of pulses. Pulses cover roughly 23% of the land dedicated to food grains in India and make up about 9-10% of the country's total food grain production. With around 30 million hectares under pulse cultivation, India produces approximately 25 million tonnes annually, achieving an average productivity of 851 kg/ha (based on data from 2018-19 to 2022-23).

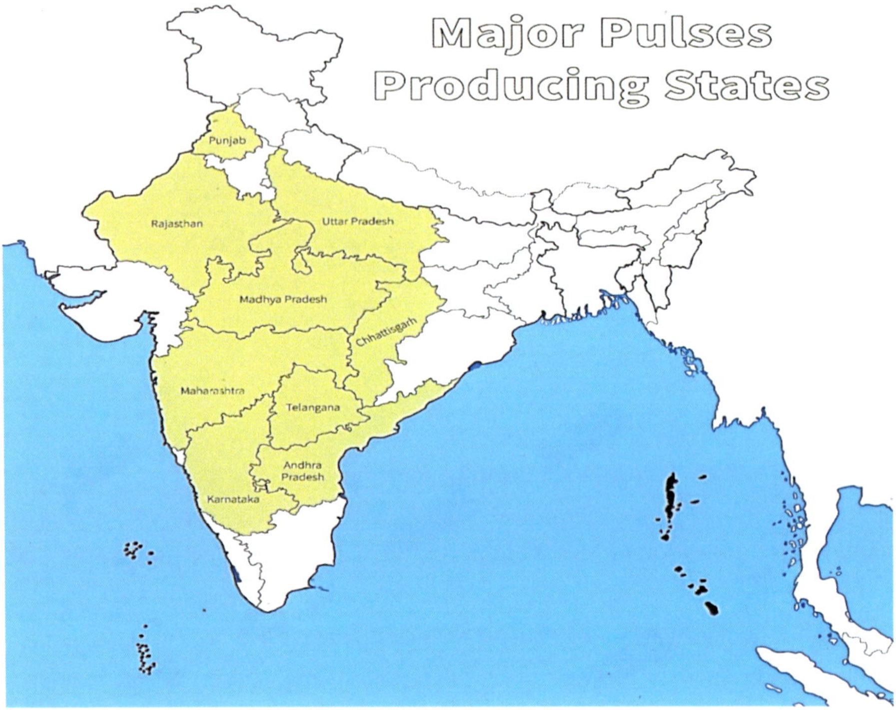

Fig. 10.2: Major pulses producing states of India

Rajasthan, Madhya Pradesh, Maharashtra, Uttar Pradesh, and Karnataka are the leading states of India in pulse production. Madhya Pradesh, Maharashtra, and Rajasthan each contribute 22, 16 and 10 respectively to the country's total pulse production. However, the current output is insufficient to meet the growing demand, leading India to import pulses. The government is focusing on increasing pulse production as part of its strategy to double farmers' incomes by encouraging a shift from traditional cereal crops. Although India accounts for 25 % of the global pulse production, its share in total food grains has declined from 16 % in 1950 to 8 % in 2022-23, resulting in a significant demand-supply gap. The study projects pulse production to reach 26 million tonnes by 2026, but to meet the demand by 2050, an annual growth rate of 2.2 % is necessary. Due to inconsistent production, the demand-supply gap is expected to widen, and the decreasing per capita availability of pulses is further exacerbated by government policies that favour wheat and rice over pulses (Sow and Ranjan, 2021).

Table 10.3: Area (lakh ha), production (lakh tonnes) and yield (kg/ha) of pulses in india

Crop	Season	Normal (Average of 2018-19 to 2022-23)		
		Area	Production	Yield
Tur	Kharif	45.55	38.11	837
Gram	Rabi	100.91	117.48	1164
Urd	Kharif	36.73	17.66	481
	Rabi	9.10	7.89	867
	Total	45.83	25.56	558
Mung	Kharif	37.01	17.61	476
	Rabi	14.12	12.17	862
	Total	51.13	29.78	583
Lentil	Rabi	14.36	13.30	926
Other Pulses	Kharif	16.73	7.59	454
	Rabi	16.61	16.03	965
	Total	33.35	23.62	708
Total Pulses	Kharif	136.02	80.97	595
	Rabi	155.10	166.87	1076
	Total	291.12	247.85	851

Source: DES, Min. of Agri. & FW (Normal 2018-19 to 2022-23)

Madhya Pradesh is the largest producer of Pulses in India followed by Maharashtra, Rajasthan, Uttar Pradesh & Karnataka as given under Table 10.4

Table 10.4: States contribution in pulses area and production

State	Area (lakh hectare)	Area contribution (%)	Production (lakh tonnes)	Yield (kg/ha)
Madhya Pradesh	54.38	19	54.82	1008
Maharashtra	45.62	16	40.80	894
Rajasthan	60.69	21	40.36	665
Uttar Pradesh	24.45	8	25.59	1046
Karnataka	31.21	11	19.45	623
Gujarat	12.07	4	16.08	1333
Andhra Pradesh	12.16	4	10.26	844
Jharkhand	7.93	3	8.23	1038
Telangana	5.31	2	5.30	998
Tamil Nadu	8.14	3	5.26	646
Others	29.14	10	21.70	851
Total Pulses	291.12		247.85	851

Source: DES, Min. of Agri. & FW (Normal 2018-19 to 2022-23)

India's Pulses Import Status

In the fiscal year 2023-24, India imported 4.65 million metric tons of pulses, up from 2.53 million tons in 2022-23, marking the highest level since 2018-19. The value of these imports surged by 93%, reaching USD 3.75 billion. Red lentil imports, particularly from Canada, doubled to 1.2 million tons. The introduction of duty-free imports in December, 2023 led to an increase in yellow pea imports from Russia and Turkey. South Asian countries, including India, typically source pulses from Canada, Myanmar, Australia, Mozambique, and Tanzania.

India's Pulses Export Status

The country exported 626,653.80 MT of pulses worldwide, valued at Rs. 5,689.40 crores (USD 686.90 million) in the year 2023-24. The primary export destinations included Bangladesh, China, the United Arab Emirates, the U.S.A., and Sri Lanka.

Table 10.5: India Export of Pulses in 2024-25 (April-June)

India Export of Pulses 2024-25 (April-June)			
Country Name	**Qty (MT)**	**Rs. Crore**	**US$**
Bangladesh	55871	360.17	43175286
China	18783	180.24	21605430
United Arab Emirates	15116	147.31	17657432
U S A	8313	122.6	14694597
Turkey	9761	115.09	13797129
Iraq	11956	99.89	11975260
Sri Lanka	7134	76.53	9172367
Canada	5268	64.72	7756254
Egypt	5003	58.56	7019861
U K	5823	55.1	6604430
Saudi Arabia	3851	42.51	5095358
Nepal	5280	37.46	4490502
Malaysia	3673	33.94	4068657
Iran	4313	31.86	3818958
Libya	2572	30.59	3667470
Morocco	1800	20.38	2443689
Singapore	1661	18.62	2232213
Djibouti	2390	16.29	1952351
Netherland	1093	12.94	1551056
Australia	883	12.69	1520544
Qatar	1218	12.25	1468297
Kuwait	847	10.41	1248495
Pakistan	991	10.22	1224753
Vietnam Soc Rep	1371	8.84	1059383

Source: APEDA statistics

India's Initiatives to Boost Pulses Production

I. **NFSM-Pulses:** The National Food Security Mission - Pulses initiative, managed by the Department of Agriculture & Farmers Welfare, is implemented across the country and the objectives are:

- Providing assistance to farmers through States/UTs for various interventions.
- Conducting cropping system demonstrations.
- Facilitating the production and distribution of high-yielding varieties (HYVs) and hybrids.
- Establishing 150 Seed Hubs for Pulses, which has greatly improved the availability of quality pulse seeds.

II. **Pradhan Mantri Annadata Aay SanraksHan Abhiyan (PM-AASHA) Scheme:** launched in 2018, this umbrella scheme includes three key components:

- **Price Support Scheme (PSS)**: Procuring pulses from pre-registered farmers at the Minimum Support Price (MSP).
- **Price Deficiency Payment Scheme (PDPS)**: Compensating farmers for the difference between market prices and the MSP.
- **Private Procurement Stockist Scheme (PPSS):** Encouraging private sector participation in the procurement process.

III. **ICAR's Role in Research and Variety Development:** The Indian Council of Agricultural Research (ICAR) plays a crucial role in enhancing the productivity of pulse crops through focused research and development. ICAR's efforts include:

- Conducting basic and strategic research on pulses.
- Collaborating with State Agricultural Universities on applied research.
- Developing location-specific high-yielding varieties and production packages.

Between 2014 and 2023, ICAR has successfully developed and recognized 343 high-yielding varieties and hybrids of pulses for commercial cultivation across the country.

Policy Imperatives

I. **Pulses Linkage to Welfare Schemes:** Improving food and nutritional security is a top priority for the government, with various welfare schemes in place to address this issue. Pulses, being a low-cost protein source, can play a significant role in enhancing the nutritional status of the population, particularly for women and children. By integrating pulses into welfare programs like the Public Distribution System (PDS), Mid-Day Meal, and Integrated Child Development Services (ICDS), the government can combat protein-energy malnutrition among vulnerable groups. This approach also benefits pulse farmers by creating a stable market, ensuring better prices, and preventing distress sales. States like Andhra Pradesh, Tamil Nadu, Himachal Pradesh, Punjab, and Chhattisgarh have already included pulses in their PDS to tackle nutritional deficiencies among the poor. Expanding this initiative nationally, through programs like the Antyodaya Anna Yojana and Targeted Public Distribution System, could further improve access to vegetarian protein for the underprivileged and support the livelihoods of pulse farmers.

II. **Building Farmers' Associations:** The marketing system for pulses in India involves numerous intermediaries, including aggregators, wholesalers, commission agents, processors, and retailers. This results in a wide distribution of market margins, significantly reducing the share of profits that pulse producers receive. Despite high market prices for pulses, farmers often do not see proportional returns due to these intermediaries. To improve this, better integration within the pulse marketing chain is needed to ensure farmers receive a fairer margin and consumers access pulses at reasonable prices. Individual farmers often lack bargaining power due to their small-scale surpluses. However, by organizing into groups or institutions like registered societies or producer organizations, farmers can pool their resources, gain collective bargaining power, and secure larger market margins. This approach would not only increase their income but also encourage better resource allocation for pulse production at the household level.

III. **Minimum Support Price (MSP):** The increase in the minimum support price and bonuses for pulse production serves as an incentive for farmers, provided the commodity is purchased at the MSP. In response to government interventions, the MSP for pulses has seen a significant rise during 2023-24, encouraging more farmers to cultivate pulses. This policy has led to a steady increase in both the area under cultivation and pulse production in India over the years. The Government of India has adopted the strategy of increasing the MSP and announcing it at the time of sowing to boost pulse production. As a result, farmers prioritize pulse cultivation, viewing it as a secure option due to these policy incentives. The assured market price through MSP can lead to greater resource allocation for pulse production. However, the MSP determination process must consider the risks and uncertainties involved in pulse production to ensure a realistic estimation of production costs. Additionally, identifying registered agencies for procurement and storage at accessible locations is crucial for enhancing farmers' incomes.

Table 10.6: Minimum Support Prices for different grain legumes

MSP (INR per quintal)					
Crop	**2021-22**	**2022-23**	**2023-24**	**Increase in MSP 2022-23 over 2021-22**	**Increase in MSP 2023-24 over 2022-23**
Tur (Arhar)	6300	6600	7000	300	400
Moong	7275	7755	8558	480	803
Urad	6300	6600	6950	300	350
Gram	5230	5335	5440	105	105
Masur (Lentil)	5500	6000	6425	500	425

Source: PIB, Cabinet Committee on Economic Affairs (CCEA), Govt. of India

Table 10.7: MSP for kharif Pulses for 2024-25

Minimum Support Prices for 2024-25 (INR per quintal)		
Crop	**MSP (2024-25)**	**MSP Increase in 2024-25 over 2023-24**
Arhar (Pigeon pea)	7550	550
Moong (Green gram)	8682	124
Urad (Black gram)	7400	450

Source: PIB, Cabinet Committee on Economic Affairs (CCEA), Govt. of India

IV. Processing of Pulses: Processing of pulses represents the largest portion of overall marketing costs. Promoting pulse processing at the village level, with strong support from stakeholders in terms of in situ processing technologies and market intelligence for identifying appropriate sales channels, can increase farmers' share of the consumer price. In India, over three-fourths of pulses are processed into "daal". However, processing losses and waste are estimated to be 12 to 16%. Reducing these losses through the use of efficient processing units can result in additional returns for pulse producers. Therefore, establishing community-based small-scale pulse milling units, particularly at the village level, would not only reduce processing losses but also promote off-farm employment and generate additional income for pulse growers.

Conclusion

The cultivation of pulses enhances agricultural biodiversity by promoting diverse cropping systems. This diversity reduces the risk of crop failure due to pests, diseases, or extreme weather events. In terms of food security, pulses are crucial for improving the nutritional status of populations, especially in regions where staple grain-based diets lack essential nutrients. Finally, the environmental benefits of pulse cultivation are substantial. Their lower water and fertilizer requirements compared to other crops, coupled with their role in nitrogen fixation, help reduce greenhouse gas emissions and support sustainable agricultural practices. Overall, pulses are indispensable in fostering a sustainable and resilient agricultural sector.

References

Abraham, M., & Pingali, P. (2021). Shortage of pulses in India: Understanding how markets incentivize supply response. Journal of Agribusiness in Developing and Emerging Economies, 11(4), 411-434.

Benitez-Alfonso, Y., Soanes, B. K., Zimba, S., Sinanaj, B., German, L., Sharma, V., et al. (2023). Enhancing climate change resilience in agricultural crops. Current Biology, 33(23), R1246-R1261.

Brockwell, J., Bottomley, P. J., & Thies, J. E. (1995). Manipulation of rhizobia microflora for improving legume productivity and soil fertility: a critical assessment. In Management of Biological Nitrogen Fixation for the Development of More Productive and Sustainable Agricultural Systems: Extended versions of papers presented at the Symposium on Biological Nitrogen Fixation for Sustainable Agriculture at the 15th Congress of Soil Science, Acapulco, Mexico, 1994, Springer Netherlands, pp. 143-180.

Gyogluu, C., Jaiswal, S. K., Kyei-Boahen, S., & Dakora, F. D. (2018). Identification and distribution of microsymbionts associated with soybean nodulation in Mozambican soils. Systematic and Applied Microbiology, 41(5), 506-515.

Hungria, M., & Mendes, I. C. (2015). Nitrogen fixation with soybean: the perfect symbiosis?. Biological Nitrogen Fixation, 1009-1024.

Lovegrove, J. A., O'Sullivan, D. M., Tosi, P., Millan, E., Todman, L. C., Bishop, J., et al. (2023). 'Raising the Pulse': the environmental, nutritional and health benefits of pulse□ enhanced foods. Nutrition Bulletin, 48(1), 134-143.

Nyaombo, J. J. (2022). Pulses Farming; An Adaptive Strategy to Climate Change in Arid and Semi-Arid Regions: A Case Study of Itigi District in Singida Region, Tanzania. In The Nature, Causes, Effects and Mitigation of Climate Change on the Environment. IntechOpen.

OECD/FAO. (2023), OECD-FAO Agricultural Outlook, OECD Agriculture statistics (database). doi:10.1787/agr-outl-data-en

Peoples, M. B., Herridge, D. F., & Ladha, J. K. (1995a). Biological nitrogen fixation: an efficient source of nitrogen for sustainable agricultural production. In Management of Biological Nitrogen Fixation for the Development of More Productive and Sustainable Agricultural Systems: Extended versions of papers presented at the Symposium on Biological Nitrogen Fixation for Sustainable Agriculture at the 15th Congress of Soil Science, Acapulco, Mexico, 1994, Springer Netherlands, pp. 3-28.

Peoples, M. B., Ladha, J. K., & Herridge, D. F. (1995b). Enhancing legume N 2 fixation through plant and soil management. In Management of Biological Nitrogen Fixation for the Development of More Productive and Sustainable Agricultural Systems: Extended versions of papers presented at the Symposium on Biological Nitrogen Fixation for Sustainable Agriculture at the 15th Congress of Soil Science, Acapulco, Mexico, 1994, Springer, pp. 83-101.

Ranjan, S., Sow, S., Kumar, S., Ghosh, M., Roy, D. K., Dutta, S. K., & Pramanick, B. (2022). Enhancing Nutrient Use Efficiency and Productivity of Cereals through Site Specific Nutrient Management. Journal of Cereal Research, 14(3), 229-242.

Rao, D. L. N., & Balachandar, D. (2017). Nitrogen inputs from biological nitrogen fixation in Indian agriculture. In The Indian nitrogen assessment, pp. 117-132.

Singh, N. (2017). Pulses: an overview. J. Food Sci. Technol. 54(4), 853-857.

Singh, N. P. (2018). Pulses as a candidate crops for doubling farmers' income. Indian Farming, 68(1).

Sow, S., & Ranjan, S. (2021). Popularization of Pulse Production for Ensuring Nutritional and Food Security. Agriculture & Food E:Newsletter, 3(2), 16-19.

Venkidasamy, B., Selvaraj, D., Nile, A. S., Ramalingam, S., Kai, G., & Nile, S. H. (2019). Indian pulses: A review on nutritional, functional and biochemical properties with future perspectives. Trends in Food Science & Technology, 88, 228-242.

Vitousek, P. M., Menge, D. N., Reed, S. C., & Cleveland, C. C. (2013). Biological nitrogen fixation: rates, patterns and ecological controls in terrestrial ecosystems. Philosophical Transactions of the Royal Society B: Biological Sciences, 368(1621).

Wani, S. P., Rupela, O. P., & Lee, K. K. (1995). Sustainable agriculture in the semi-arid tropics through biological nitrogen fixation in grain legumes. In Management of Biological Nitrogen Fixation for the Development of More Productive and Sustainable Agricultural Systems: Extended versions of papers presented at the Symposium on Biological Nitrogen Fixation for Sustainable Agriculture at the 15th Congress of Soil Science, Acapulco, Mexico, 1994, Springer Netherlands, pp. 29-49.

Zahran, H. H. (1999). Rhizobium-legume symbiosis and nitrogen fixation under severe conditions and in an arid climate. Microbiology and Molecular Biology Reviews, 63(4), 968-989.

Zhao, J., Chen, J., Beillouin, D., Lambers, H., Yang, Y., Smith, P., et al. (2022). Global systematic review with meta-analysis reveals yield advantage of legume-based rotations and its drivers. Nature Communications, 13(1), 4926.

11

Nutrient Cycling and Extraction Pattern in Soil Under Grain Legume Cultivation

K. Kiruthika*[*]*and R. Sowmiya

Department of Agronomy, Vanavarayar Institute of Agriculture, Pollachi, Tamil Nadu

[*]Corresponding author's email: kiruthikaagri98@gmail.com

Abstract

The cultivation of grain legumes is expanding globally because of their diverse applications as food for humans, feed for livestock, and raw materials for industrial purposes. Legumes contribute significantly to sustainable agricultural practices due to their unique ability to fix atmospheric nitrogen through symbiosis with rhizobia bacteria in their root nodules. Incorporating legumes into cropping systems in place of traditional cereal-cereal rotations can significantly enhance agricultural sustainability due to their multiple benefits. Their role in nutrient cycling, particularly nitrogen and phosphorus, ensures that subsequent crops benefit from improved soil nutrient availability, making legumes an essential component of crop rotations and intercropping systems. Additionally, legumes improve soil structure, organic matter content, and decrease the environmental impact associated with excessive fertilizer use, such as greenhouse gas emissions and water pollution. With the increasing demand for plant-based foods and the economic challenges of fertilizing cereal crops, grain legumes are anticipated to play a more prominent role in cropping systems, especially in regions where fertilizer availability and affordability are limited. It includes studies on improving nitrogen fixation efficiency, enhancing phosphorus uptake, and developing varieties that are better adapted to specific environmental conditions and cropping systems. Additionally, there is a need to identify the most suitable legume species and varieties for different regions and cropping patterns to ensure maximum productivity and sustainability.

Keywords: *Climate change mitigation, Grain legumes, N fixation, P availability, Soil organic carbon*

Introduction

Legumes called as poor man's meat, play a significant role in human nutrition, particularly for low-income populations in developing nations. The family

Leguminosae is the third largest among flowering plants, comprising 20,000 species and 800 genera. Grain legumes provide approximately one-third of the plant protein and a similar share of the vegetable oil consumed by humans. The protein content in grain legumes varies from 20% in common beans and lentils to 40% in soybeans and yellow lupins, which is higher compared to cereals (7–17%) and rapeseed (17–26%) as noted by Day (2013). They are also good sources of slow-release carbohydrates (dietary fiber), vitamins and minerals. Various types of legumes, including red gram (pigeon pea, *Cajanus cajan*), black gram (*Vigna mungo* L.), broad bean (*Vicia faba* L.), Bengal gram (chickpea, *Cicer arietinum* L.), cowpea (*Vigna unguiculata* L.), field bean (*Dolichos lablab*), green gram (*Vigna radiata*), horse gram (*Macrotyloma uniflorum*), and so on, are commonly consumed in India, mostly by the weaker sections of the population. Legumes are known to provide numerous physiological benefits that help to regulate and prevent a variety of metabolic disorders, including diabetes mellitus, coronary heart disease, and colon cancer (Tharanathan & Mahadevamma, 2003). Increasing domestic production of legumes is beneficial as it supports sustainable development through various means, such as reducing reliance on fossil fuels in agriculture, lowering greenhouse gas (GHG) emissions, enhancing crop diversity within cropping systems, boosting above and below-ground biodiversity, improving soil fertility, increasing carbon storage, and strengthening the connection between crop and livestock production (Ranjan *et al.*, 2022). Understanding the nutrient cycling and extraction patterns in soil under grain legume cultivation is essential for optimizing crop productivity, maintaining soil health, and promoting sustainable agriculture. This chapter explores the mechanisms of nutrient cycling and the role of legumes in soil nutrient dynamics.

Legumes in agricultural system

A decline in soil nutrient levels is a widespread issue, posing a fundamental challenge to food production. The growing emphasis on low-input crop production and sustainable agricultural systems from environmental, economic, and management perspectives has paved the way for the continued adoption and integration of legumes into agricultural practices.

The inclusion of legumes in the crop production system has several advantages:

i) At the food system level, legumes play a vital role in the food chain for both human and animal consumption. They are the major source of plant proteins which is important for enhancing human health.

ii) At the production level, their ability to fix atmospheric nitrogen which is an inexpensive, environmentally friendly N fertilizer.

iii) At the cropping system level, grain legumes can be included as diversification crops in rotation with cereals and oilseeds which offers multiple benefits jeopardizing the pest and disease cycle (Voisin *et al.*, 2014).

Thus, legumes play a significant role in sustaining global food security, maintaining the ecosystem, optimizing resource use efficiency, and contributing to the balance of plant protein production.

Biological N fixation by legumes

Legumes form a symbiotic relationship with rhizobia bacteria, which convert inert atmospheric nitrogen (N) into a usable form within the plant's root nodules through biological nitrogen fixation (BNF). This process is a key natural source of nitrogen cycled for sustaining ecosystems. Symbiotic nitrogen fixation in legumes contributes approximately 8.0 × 10 kg of nitrogen, with an average fixation rate ranging from 20 to 200 kg of N ha^{-1} $year^{-1}$ (Vasconcelos *et al.*, 2020; Iannetta *et al.*, 2016). The BNF will vary for legume species and it depends on the various characteristics of soil, climatic and cropping systems followed. (i.e., monoculture, mixed culture, crop rotations, etc.) as well as soil management techniques. Symbiotically fixed N_2 is very important in sustainable agriculture since it enables a reduction in the amount of costly chemical N used in crop production. In addition to supplying nitrogen through symbiosis, legumes facilitate the solubilization of unavailable phosphorus (P) in the soil, enhance microbial activity, improve the soil physical structure, increase organic matter, and suppress weed growth.

Table 11.1: BNF capacity (kg ha^{-1}) of commonly cultivated grain legumes (Vasconcelos *et al.*, 2020)

Common name	Scientific name	BNF (kg ha^{-1})	Cultivation area
Pea	*Pisum sativum* L.	36.6–125.3	Canada, Greece
Chickpea	*Cicer arietinum* L.	21.0–103.6	Canada
Lentil	*Lens culinaris* Med.	23.0–86.8	Switzerland, Canada
Common bean	*Phaseolus vulgaris* L.	16.3–71.9	Canada
Cowpea	*Vigna unguiculata* (L.)	36–75	Brazil

Nutrient cycling and extraction pattern in legume-based system

Direct Nitrogen Transfer

Direct nutrient transfer involves the movement of nutrients from a nitrogen-fixing legume to another crop during its growth within an intercropping system, or as residual nitrogen that benefits a subsequent crop. In mixed farming systems, legumes aid in better utilization of soil nutrients by the release and fixed N during the cereal phase, leading to better N yield of cereal (Sow *et al.*, 2023). The nitrogen transfer depends on multiple factors in which soil physical, chemical and biological properties have a greater role. The biotic factors such as root contact, plant density, growth stage, production year, defoliation, and root herbivores, as well as abiotic factors like water stress, temperature, light, and soil nitrogen availability (Thilakarathna *et al.,* 2016) affects N transfer.

The belowground nitrogen transfer occurs through distinct pathways which include

a) **Decomposition:** The neighbor plants and the subsequent crops uptake N released from the decomposition of legumes.

b) **Root exudation:** The exudation of soluble nitrogen compounds by legumes plants, which are then absorbed by nearby plants.

c) **Mycorrhizae:** The transfer of N by plant-associated mycorrhizae.

Long-term nitrogen transfer mainly happens through the decomposition of roots and nodules, while short-term transfer is driven by nitrogen-containing root exudates and mycorrhizae.

a) **Nitrogen transfer through decomposition of roots and nodules:** In legumes, the part of symbiotically fixed N becomes available to subsequent crops through the decomposition and mineralization of legume residues. Legume residues often provide more mineral nitrogen to succeeding crops compared to cereal residues, due to their higher nitrogen content and lower C ratio. Nitrogen derived from decomposed roots, nodules, root caps, root border cells, sloughed cells, and the epidermis (water-insoluble materials) plays a significant role in belowground nitrogen transfer (Wichern *et al.,* 2008). When comparing different mechanisms of nitrogen transfer, the decomposition of nodules and roots is considered more important than nitrogen transfer through root exudates or mycorrhizae. Although the decomposition of belowground plant tissues is a key contributor to nitrogen transfer, it is typically a slower process compared to nitrogen transfer via root exudates and mycorrhizae. It is estimated that nitrogen transfer through the decomposition of legume nodules and roots ranges from 3 to 102 kg ha^{-1} yr^{-1}, accounting for 2–26 % of the biologically fixed nitrogen in legumes. Nitrogen derived from the decomposition of roots and nodules primarily contributes to the later stages of plant growth or to production in subsequent years.

b) **Nitrogen transfer by root exudates:** Root exudates, the compounds released by the root system, can be either low-molecular-weight (e.g., amino acids, sugars) or high-molecular-weight (e.g., proteins). During early growth stage of legume, most nitrogen transfer occurs through exudates rather than root or nodule decomposition. Plants can absorb organic nitrogen from these exudates, enabling direct nitrogen transfer to non-legumes (Fig. 1). Additionally, due to their low C : N ratio, exudates are quickly mineralized by soil microbes, promoting indirect nitrogen transfer (Lesuffleur and Cliquet, 2010). Ammonium and amino acids are the main nitrogen compounds exuded by most temperate legumes, primarily through root nodules and tips. It is estimated that nitrogen derived from legume rhizodeposits increases residual soil nitrogen by 35–44% and makes up 79–85% of below-ground nitrogen at plant maturity (Mayer *et al.,* 2003) and 47% of below-ground nitrogen originates from legume root deposits. The type of nitrogen exudate varies by legume; temperate legumes like alfalfa mainly release amino-N or NH_4-N, whereas tropical legumes like soybeans primarily exude ureides.

c) **Nitrogen Transfer Mediated by Mycorrhizae:** Arbuscular mycorrhizal fungi (AMF) enhance nitrogen transfer between plants by forming mycelial connections or increasing capacity of receiver plants to uptake legume derived N by increasing the soil volume. AMF can absorb NH_4^+, NO_3^-, and organic nitrogen, efficiently scavenging nitrogen from decomposing organic material. In legume/non-legume mixed-cropping systems, AMF capture nitrogen released by legumes and transfer it to nearby non-legumes (Jannoura *et al.,* 2012) (Fig. 1). While nitrogen transfer through AMF can be bidirectional, the majority (up to 80%) moves from nitrogen-fixing legumes to non-fixing plants, with less than 10% transferring in the opposite direction.

Role of Grain legumes in maintaining nutrient balance in Soil

- Nitrogen Fixation
- Improving Soil Structure and Organic Matter
- Increasing Phosphorus availability
- Enhancing Soil bio-Pores
- Reducing Nutrient Leaching

- Balancing Soil pH
- Breaking Disease and Pest Cycles
- Supporting Beneficial Microorganisms

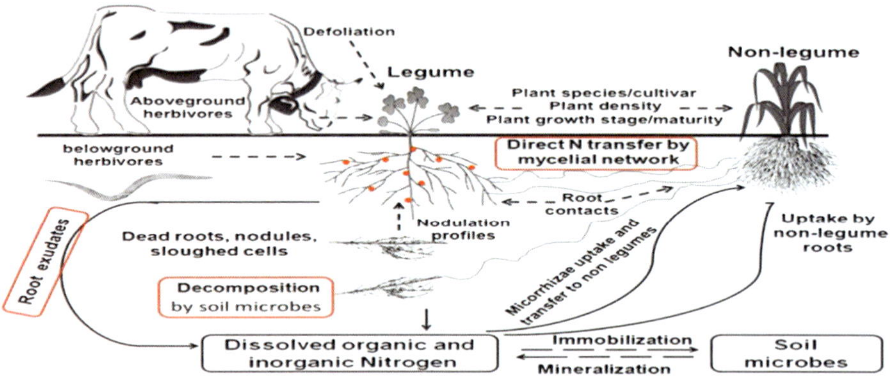

Fig. 11.1: Below-ground transfer of nitrogen from legumes to non-legume plants (Thilakarathna *et al.*, 2016)

i) **Contribution of Legumes to Soil P uptake:** Legume-based systems enhance soil fertility including SOC and humus content, as well as N and P availability. Phosphorus, an essential plant nutrient is highly immobile in soils since it is adsorbed in soil surfaces. Intercropping with legumes found to increase the availability of phosphorous in the rhizosphere region (Murphy-Bokern *et al.*, 2017). The soil type, environment and climatic factors greatly decide the P solubilization. Legume roots exude organic acids which lower the pH in the surrounding soil, thereby solubilizing and releasing phosphorus that is in unavailable form. Grain legumes can also release phosphatase enzymes into the soil to break down organic material containing phosphorus and a direct interaction between the root surface and the insoluble phosphorus near the root increase P uptake.

ii) **Contribution of Legumes to SOC:** Legume-based cropping systems increases the amount of crop's residual carbon (C) that is stored in the soil as stable Soil Organic Matter (SOM). Grain legumes increase Soil Organic Carbon (SOC) by producing biomass, organic C, and N as well as releasing hydrogen gas as a by-product of BNF, which promotes bacterial legume nodules' development in the rhizosphere (Stagnari *et al.*, 2017). Borase *et al.* (2021) reported that adding mung bean to a Rice-Wheat rotation increased the soil microbial activity and SOC level which ultimately resulted in soil aggregation and carbon sequestration.

This might be due to extensive ground cover in all three seasons in the Rice-Wheat-Mung bean rotation.

A study on the effect of grain legumes in sandy soil after three years of research registered that higher content of organic carbon after legume harvest as compared to soils with oats grown (7.21 $g \cdot kg^{-1}$ DM, on average) (Hajduk *et al.*, 2015). Legume-inclusive crop rotations with Integrated Nutrient Management (INM) positively affect soil properties, especially when legume residues are returned to the soil. For example, rotations Banerjee *et al.*, (2016) like rice-chickpea (R-C) and rice-wheat-rice-chickpea (R-W-R-C) showed higher SOC than rice-wheat (R-W) rotations. This is partly due to legume roots, which are resistant to decomposition, and the higher release of root exudates from grain legumes, contributing to more stable organic matter in the soil.

iii) **Enhancement of Soil Microorganisms:** During biological nitrogen fixation (BNF), hydrogen gas is produced, which promotes bacterial growth in the legume rhizosphere, resulting in increased microbial biomass carbon in the soil. Alvey *et al.* (2003) reported that incorporating legume crop rotations significantly impacts the microbial community structure and enhances microbial diversity. The high protein content of legumes supports the decomposition of crop residues by stimulating microbial growth in the soil.

iv) **Enhancement of Soil Biopores:** Nitrogen-rich legume residues enhance the formation of soil bio-pores both directly through the decomposition of residues and indirectly by boosting earthworm populations in the rhizosphere.

v) **Prevention of Nutrient Leaching:** Incorporating legumes into the soil helps prevent the leaching of essential nutrients, reduces the adverse effects of harmful agro-chemicals, and mitigates the impact of soil-borne pathogens (Ochieng *et al.*, 2023).

vi) **Breaking Disease and Pest Cycles:** Including legumes in crop rotations helps break the cycles of pests and diseases commonly associated with continuous cereal cropping. This reduces the need for chemical inputs and maintains a healthier nutrient balance in the soil.

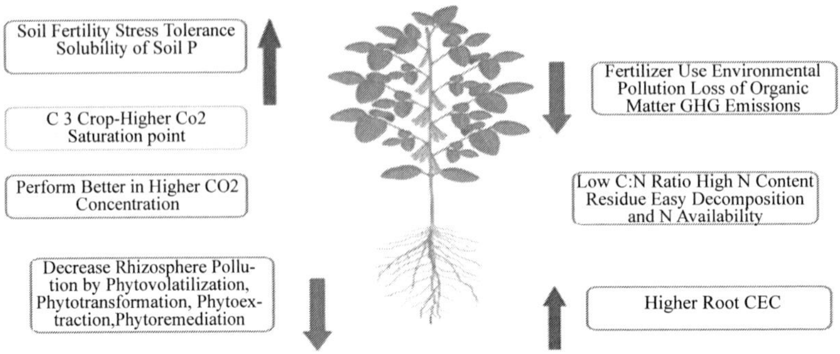

Fig. 11.2: Importance and role of grain legumes

Benefits of Grain Legumes in Cropping System

Legumes can be integrated into cropping systems such as intercropping, sequential cropping, crop rotation, improved fallows, green manuring and alley cropping. The addition of grain legumes into Rice Wheat rotation enhances soil health, crop productivity, and resource use efficiency. The incorporation of legumes into cropping systems results in improved soil fertility and higher crop yields. The rice (base-crop) grain yield was increased by 15% ($p < 0.05$) when mung bean was added to the rice-wheat rotation. Likewise, the addition of chickpea (R-C) increased rice crop production by 7%.

Crop rotations with legumes improve the soil fertility and soil structure, increase microbial activity, organic matter levels, water retention and provide resistance to erosion. Furthermore, when grain legumes are included in crop rotations, they enhance soil nitrogen levels, benefiting the non-N-fixing crops that follow (Table 2). Increasing the proportion of legume crops within a rotation has the potential to significantly reduce the dependency on inorganic fertilizers. This practice not only enhances the contribution of renewable nitrogen to global nutrient cycles but also may lead to a decrease in the amount of reactive nitrogen lost from ecosystems. Legume-based cropping systems significantly reduce the N losses and nutrient leaching. Grain legumes as intercrops with cereals resulted in higher N output and reduced nitrate leaching reported by Hauggaard-Nielsen *et al.* (2009); Reckling *et al.,* (2014). In regions with higher intensity of leaching or denitrification, legume-based cropping systems achieved more N synchrony than conventionally managed fertilizer-based systems. Similarly, in lowland rice and irrigated systems, the plant recovery of N was found to be greater and the total N loss was less compared with fertilizer (Crews and Peoples, 2005). Nitrogen-rich legume residues enhance the formation of soil bio-pores both directly through the decomposition of residues and indirectly by boosting earthworm populations in the rhizosphere. This process helps optimize the balance of air and water in the crop root zone. Thus, adding legumes in intercropping or crop rotation is considered to be a sustainable and alternate method of adding nitrogen to low-input cropping systems.

Table 11.2: Nitrogen fixed and N Transfer to soil and subsequent crops in Various agricultural systems (Anowarul Islam and Adjesiwor, 2017)

Crop(s)	% of fixed nitrogen	Amount of N transferred
Mung bean–oat	9.7	12.8 mg N plant^{-1}
Soybean–maize	7.57	7.84 mg N pot^{-1}
Soybean–maize	1.26 – 2.17	10.77–13.72 mg N pot^{-1}
Faba bean–wheat	14.9	0.17 mg N plant shoot^{-1}
Pigeon pea (L.) - coffee	NA	21.8 g N kg^{-1}

NA- data not available/ estimated

Effect of Grain Legumes on Atmosphere and Soil Quality

Legumes provide numerous advantages, yet their potential to mitigate climate change has been rarely addressed. Legumes can reduce the emission of GHGs like carbon dioxide (CO_2) and nitrous oxide (N_2O). They also have an important role in soil carbon sequestration and reduction of overall fossil energy inputs (Jensen *et al.,* 2012). Agriculture accounts for 60% of anthropogenic N_2O emissions from both animal and crop production (Reay *et al.,* 2012). The application of nitrogen fertilizers is the primary cause of most of these emissions; for every 100 kg of N fertilizer, approximately 1.0 kilogram of N is released as N_2O, while actual amounts vary depending on several factors such as soil pH, texture, organic C content, and rate of N application.

Recent studies have concentrated on how legumes can help reduce greenhouse gas emissions. Compared to other crops, legume crops release five to seven times less GHG per unit area. By examining N_2O fluxes, they discovered that peas released 69 kg N_2O ha^{-1}, a significantly lower amount than rape (534 kg N_2O ha^{-1}) and winter wheat (368 kg N_2O ha^{-1}). The inclusion of legumes into agricultural rotations helps to reduce the usage of fertilizers and energy in arable land which in turn reduces GHG emissions. In rotations with leguminous crops, the annual savings from N fertilizer in Europe range from 277 kg ha^{-1} of CO_2 (1-kilogram N = 3.15 kg CO_2) (Stagnari *et al.,* 2017).

Nitrous oxide (N_2O), a greenhouse gas, is becoming more concentrated in the atmosphere, mostly due to an increase in anthropogenic nitrogen (N) inputs into the soil from crop and animal production. The biological processes of nitrification and denitrification, which utilize inorganic N are the source of soil emissions of N_2O. (Schwenke *et al.,* 2015) In subtropical rainfed cropping systems, raising legume crops greatly reduces the amount of nitrous oxide emission. The impact of the N source on total emissions is probably caused by the different rates at which mineral N is available from fertilizers (fast-release) and legumes (slow-release). Also, the timing of mineral N release from fertilizer and legume-derived N accounts for the difference. Lemke *et al.* (2007) also documented that the pulse phase of crop rotations produces comparatively less N_2O emissions than the fertilized cereal in the Northern Great Plains. Similarly, CO_2 emissions were found to be lower with the inclusion of pulse crops than non-pulse crop rotation. High biomass input legume cover crops along with conservation management practices like no-tillage practices could result in higher soil CO_2 sequestration. The increased biomass content of legumes and less organic matter mineralization functions as a sink of greenhouse gases (GHG), especially in moist and hot subtropical regions, where the environmental conditions are very conducive to microbial activity and fluxes of C- and N-derived gases (Bayer *et al.*, 2016).

Table 11.3: Examples of the fate of ^{15}N-labeled fertilizers or legume residues applied to field experiments in different regions of the world

Crop and system	N source	Region	N added (kg ha^{-1})	Time from N application to sowing	Time from N application to harvest	%N in crop	%N in soil	%N lost	References
Rainfed									
Corn	Urea	Nigeria	-	-	-	27	15	58	(Vanlauwe *et al.*, 2001)
	Leucaena		-	-	-	9	56	35	(Vanlauwe *et al.*, 1998)
Corn (alley crop)	Urea	Indonesia	60	1 month	70 days	22 (11)	33	34	(Rowe *et al.*, 2005; Rowe *et al.*, 2001)
	Gliricidia		178	1 month	70 days	6 (9)	58	27	Rowe *et al.*, 2001)
Wheat	$(NH_4)_2SO_4$	Canada (3 sites/2 years)	50	9 months	13-14 months	16-33	33-38	29-51	(Janzen *et al.*, 1990)
	Lentil		85	9 months	13-14 months	12-27	48-59	14-40	
Barley	NH4NO3	Sweden	80	0	3 months	36	-	-	(Bergström and Kirchmann, 2004)
	Red clover		160	0	3 months	15			
Irrigated									
Cotton	Urea	Australia	100	9 months	14 months	17		83	(Rochester *et al.*, 2001)
	Faba bean		100	9 months	14 months	62		38	
	Field pea		100	9 months	14 months	80		20	
Cotton	Urea	Australia	100	5-6 months	10-11 months	4		96	(Rochester *et al.*, 2001)
	Soyabean		100	5-6 months	10-11 months	66		36	

Conclusion

The advantages of legumes in cropping systems are elucidated through mechanisms such as direct nitrogen transfer, the residual effect of fixed nitrogen, enhanced nutrient bioavailability and uptake, improved soil physicochemical properties, disruption of pest and disease cycles, and the stimulation of soil microbial communities. The diverse potential roles of grain legumes in cropping systems must be recognized as a key focus for future research. There is a growing need to expand research efforts on grain legumes to optimize their productivity, particularly in terms of nitrogen and phosphorus accumulation, as well as enhancing both seed production and the return of vegetative residues to the cropping system.

References

Alvey, S., Yang, C.-H., Bürkert, A., & Crowley, D. (2003). Cereal/legume rotation effects on rhizosphere bacterial community structure in West African soils. Biology and fertility of soils, 37, 73-82.

Anowarul Islam, M., & Adjesiwor, A. T. (2017). Nitrogen Fixation and Transfer in Agricultural Production Systems. Nitrogen in Agriculture-Updates. doi:10.5772/intechopen.71766

Bayer, C., Gomes, J., Zanatta, J. A., Vieira, F. C. B., & Dieckow, J. (2016). Mitigating greenhouse gas emissions from a subtropical Ultisol by using long-term no-tillage in combination with legume cover crops. Soil and Tillage Research, 161, 86-94.

Bergström, L., & Kirchmann, H. (2004). Leaching and crop uptake of nitrogen from nitrogen-15-labeled green manures and ammonium nitrate. Journal of Environmental Quality, 33(5), 1786-1792.

Borase, D. N., Murugeasn, S., Nath, C. P., Hazra, K. K., Singh, S. S., Kumar, N., et al. (2021). Long-term impact of grain legumes and nutrient management practices on soil microbial activity and biochemical properties. Archives of Agronomy and Soil Science, 67(14), 2015-2032.

Crews, T. E., & Peoples, M. B. (2005). Can the synchrony of nitrogen supply and crop demand be improved in legume and fertilizer-based agroecosystems? A review. Nutrient Cycling in Agroecosystems, 72, 101-120.

Day, L. (2013). Proteins from land plants–potential resources for human nutrition and food security. Trends in Food Science & Technology, 32(1), 25-42.

Hajduk, E., Właśniewski, S., & Szpunar-Krok, E. (2015). Influence of legume crops on content of organic carbon in sandy soil. Soil Science Annual, 66(2), 52-56.

Hauggaard-Nielsen, H., Mundus, S., & Jensen, E. S. (2009). Nitrogen dynamics following grain legumes and subsequent catch crops and the effects on succeeding cereal crops. Nutrient Cycling in Agroecosystems, 84, 281-291.

Iannetta, P. P., Young, M., Bachinger, J., Bergkvist, G., Doltra, J., Lopez-Bellido, R. J., et al. (2016). A comparative nitrogen balance and productivity analysis of legume and non-legume supported cropping systems: the potential role of biological nitrogen fixation. Frontiers in Plant Science, 7, 1700.

Jannoura, R., Kleikamp, B., Dyckmans, J., & Joergensen, R. G. (2012). Impact of pea growth and arbuscular mycorrhizal fungi on the decomposition of ^{15}N-labeled maize residues. Biology and Fertility of Soils, 48, 547-560.

Janzen, H., Bole, J., Biederbeck, V., & Slinkard, A. (1990). Fate of N applied as green manure or ammonium fertilizer to soil subsequently cropped with spring wheat at three sites in western Canada. Canadian Journal of Soil Science, 70(3), 313-323.

Jensen, E. S., Peoples, M. B., Boddey, R. M., Gresshoff, P. M., Hauggaard-Nielsen, H., JR Alves, B., & Morrison, M. J. (2012). Legumes for mitigation of climate change and the provision of feedstock for biofuels and biorefineries. A review. Agronomy for Sustainable Development, 32, 329-364.

Lemke, R., Zhong, Z., Campbell, C., & Zentner, R. (2007). Can pulse crops play a role in mitigating greenhouse gases from North American agriculture? Agronomy Journal, 99(6), 1719-1725.

Lesuffleur, F., & Cliquet, J.-B. (2010). Characterisation of root amino acid exudation in white clover (Trifolium repens L.). Plant and Soil, 333, 191-201.

Mayer, J., Buegger, F., Jensen, E. S., Schloter, M., & Heß, J. (2003). Estimating N rhizodeposition of grain legumes using a 15N in situ stem labelling method. Soil Biology and Biochemistry, 35(1), 21-28.

Murphy-Bokern, D., Stoddard, F. L., & Watson, C. A. (2017). Legumes in cropping systems: CABI, pp 37-54.

Ochieng, I. O., Ranjan, S., Seleiman, M. F., Padhan, S. R., Psiwa, R., Sow, S., et al. (2023). Increasing rainwater use efficiency, gross return, and grain protein of rain-fed maize under nitrate and urea nitrogen forms. Notulae Botanicae Horti Agrobotanici Cluj-Napoca, 51(3), 13293-13293.

Ranjan, S., Sow, S., Kumar, S., Ghosh, M., Roy, D. K., Dutta, S. K., & Pramanick, B. (2022). Enhancing Nutrient Use Efficiency and Productivity of Cereals through Site Specific Nutrient Management. Journal of Cereal Research, 14(3), 229-242.

Reay, D. S., Davidson, E. A., Smith, K. A., Smith, P., Melillo, J. M., Dentener, F., & Crutzen, P. J. (2012). Global agriculture and nitrous oxide emissions. Nature Climate change, 2(6), 410-416.

Reckling, M., Schläfke, N., Hecker, J.-M., Bachinger, J., Zander, P., Bergkvist, G., et al. (2014). Legume Futures Report 4.2 Generation and evaluation of legume-supported crop rotations in five case study regions across Europe.

Rochester, I., Peoples, M., Hulugalle, N., Gault, R., & Constable, G. (2001). Using legumes to enhance nitrogen fertility and improve soil condition in cotton cropping systems. Field Crops research, 70(1), 27-41.

Rowe, E. C., Noordwijk, M. V., Suprayogo, D., & Cadisch, G. (2005). Nitrogen use efficiency of monoculture and hedgerow intercropping in the humid tropics. Plant and Soil, 268, 61-74.

Rowe, E. C., van Noordwijk, M., Suprayogo, D., Hairiah, K., Giller, K. E., & Cadisch, G. (2001). Root distributions partially explain 15N uptake patterns in Gliricidia and Peltophorum hedgerow intercropping systems. Plant and Soil, 235, 167-179.

Schwenke, G. D., Herridge, D. F., Scheer, C., Rowlings, D. W., Haigh, B. M., & McMullen, K. G. (2015). Soil N2O emissions under N2-fixing legumes and N-fertilised canola: a reappraisal of emissions factor calculations. Agriculture, Ecosystems & Environment, 202, 232-242.

Sow, S., Singh, G., Ghosh, M., Dutta, S. K., Mandal, N., Kumar, S., & Ranjan, S. (2023). Nitrogen-management strategy through leaf-colour chart and SPAD meter for optimizing the productivity in irrigated wheat (Triticum aestivum). Indian Journal of Agronomy, 68(2), 219-222.

Stagnari, F., Maggio, A., Galieni, A., & Pisante, M. (2017). Multiple benefits of legumes for agriculture sustainability: an overview. Chemical and Biological Technologies in Agriculture, 4, 1-13.

Tharanathan, R., & Mahadevamma, S. (2003). Grain legumes—a boon to human nutrition. Trends in Food Science & Technology, 14(12), 507-518.

Thilakarathna, M. S., McElroy, M. S., Chapagain, T., Papadopoulos, Y. A., & Raizada, M. N. (2016). Belowground nitrogen transfer from legumes to non-legumes under managed herbaceous cropping systems. A review. Agronomy for Sustainable Development, 36, 1-16.

Vaconcelos, M. W., Grusak, M. A., Pinto, E., Gomes, A., Ferreira, H., Balázs, B., et al. (2020). The biology of legumes and their agronomic, economic, and social impact. The plant family Fabaceae: biology and physiological responses to environmental stresses, 3-25.

Vanlauwe, B., Sanginga, N., & Merckx, R. (1998). Recovery of Leucaena and Dactyladenia residue nitrogen-15 in alley cropping systems. Soil Science Society of America Journal, 62(2), 454-460.

Vanlauwe, B., Sanginga, N., & Merckx, R. (2001). Alley cropping with Senna siamea in South-western Nigeria: I. Recovery of 15N labeled urea by the alley cropping system. Plant and Soil, 231, 187-199.

Voisin, A.-S., Guéguen, J., Huyghe, C., Jeuffroy, M.-H., Magrini, M.-B., Meynard, J.-M., et al. (2014). Legumes for feed, food, biomaterials and bioenergy in Europe: a review. Agronomy for Sustainable Development, 34, 361-380.

Wichern, F., Eberhardt, E., Mayer, J., Joergensen, R. G., & Müller, T. (2008). Nitrogen rhizodeposition in agricultural crops: methods, estimates and future prospects. Soil Biology and Biochemistry, 40(1), 30-48.

12

Reducing Carbon Footprints Role of Legumes in Low-Emission Farming Systems

Inayat M. Khan*, Aneesa Manzoor, Asma Shakeel and Andleeba Jan

Division of Soil Science & Agricultural Chemistry, FoA, Wadura, SKUAST-Kashmir India

**Corresponding author's email: khan_inayat@rediffmail.com*

Abstract

The growing global population has intensified farming practices, increasing the use of fossil fuels and agrochemicals. With limited land and the need to enhance food production, balancing environmental impact, especially carbon footprints, is crucial. Climate-smart mitigation options, with a prime emphasis on crop diversification, can help agriculture adapt to anomalous weather patterns and build resilience. Legumes offer a sustainable solution by reducing greenhouse gas emissions, as they produce 5-7 times less gases per unit area compared to other crops. They aid in carbon sequestration, with soil carbon estimates of 7.21 g kg^{-1} of dry matter, and save fossil energy by decreasing nitrogen fertilizer use, reducing CO_2 emissions by 277 kg ha^{-1} per year. Legumes play a key role in low-emission farming systems due to their ability to fix atmospheric nitrogen, reducing dependence on energy-intensive synthetic fertilizers. They improve soil health, enhance soil structure, and support carbon sequestration. Integrating legumes into crop rotations reduces the need for chemical pesticides and fertilizers. Additionally, legumes can be used in food, feed, and biofuels, lowering reliance on more carbon-intensive resources. Embracing legumes in farming systems promotes sustainability improves soil health, and considerably reduces the overall environmental impact of agriculture.

Keywords: *Carbon Footprint, Greenhouse Gas Emissions, Legumes Sustainability*

Introduction

The global population is projected to strike 9.6 billion people by 2050 (Yanni *et al.*, 2023), engraving the challenges of food security, changing climate risks due to a drop in the net discharge of greenhouse emanations into the atmosphere, and fulfilling the escalating call for energy. Specifically, the imprints of climate change and affiliated biotic and abiotic stresses that crop systems will progressively face pose grave implications for global food production (Yadav *et al.*, 2015). Food security is considered a fundamental human right essential for human development. However, ensuring food security is a multifaceted endeavor encompassing political, economic, environmental, social, and cultural dimensions. It is estimated that over 820 million people worldwide do not have access to enough food, and many others consume unhealthy diets that lead to premature death and illness (Willett *et al.*, 2019). Hence, it is crucial to optimize food production sustainably to reduce hunger, advance life expectancy, decrease infant and child mortality rates, and alleviate global poverty (Ferreira *et al.*, 2021).

The ongoing increase in greenhouse gas (GHG) concentrations in the atmosphere is a major global concern (Chai *et al.*, 2019). In 2019, global greenhouse gas (GHG) emissions amounted to 106 billion tons of carbon dioxide equivalent (CO_2-eq), with the agricultural food production system accounting for 17 billion tons (16%) (Tubiello *et al.*, 2021). Agriculture stands as one of the most susceptible sectors to imprints of climate change and one of the considerable springs of carbon emanations (Wang and Zhang, 2023). Food systems account for 21-37 % of greenhouse gas (GHG) emanations, with livestock representing the major source (Mirzabaev *et al.*, 2023). Livestock production is a significant source of the three main greenhouse gases (CO_2, CH_4, N_2O), contributing to 14.5% of total anthropogenic GHG emissions. Food systems are responsible for at least one-third or more global GHGs (Grossi *et al.*, 2019). There is a significant disparity in greenhouse gas emissions between animal-based and plant-based foods. It has been reported that ruminant meat (beef and lamb) emissions per gram of protein are approximately 250 times higher than those of legumes (Tilman and Clark, 2014). Life cycle assessment research has illustrated that pork, chicken, and seafood generate fewer greenhouse gas emissions in contrast to beef. However, even the animal products possessing the lowest imprints outdo the normal greenhouse gas emissions of plant products (Poore and Nemecek, 2018). The manufacture and employment of mineral fertilizers are imperative to ensure an adequate supply of plant nutrients for sustainable food production and the mounting populace (Brentrup *et al.*, 2018). The production of mineral fertilizers, particularly nitrogen (N) fertilizers, plays a significant role in the carbon footprint of crops as well as food production. While nitrogen fertilizers are crucial for enhancing agricultural productivity, they are also a major source of nitrous oxide (N_2O), a greenhouse gas (GHG) with a global warming potential approximately 300 times greater than that of carbon dioxide (CO_2) (Özkose, 2024). The production and application of nitrogen fertilizers contribute to roughly 5% of global GHG emissions (Gao and Cabrera Serrenho, 2023).

Climate-smart mitigation options, with a prime emphasis on crop diversification, can help agriculture adapt to anomalous weather patterns and build resilience. Legumes play a vital role in establishing and maintaining the natural ecosystem, agriculture, and agroforestry by fixing atmospheric nitrogen. Additionally, they have been a valuable source of protein in the human diet for centuries. Recently, legumes have garnered attention not only as ecologically sustainable food source rich in protein but also for their health benefits in relation to various chronic non-communicable diseases. Their positive health effects have been well-documented when included as part of many dietary patterns for promoting healthy eating and managing diseases, encircling the Dietary Approaches to Stop Hypertension (DASH) (Theodoridis *et al.*, 2023), Mediterranean diet (Rees *et al.*, 2019), high fiber diets (Kelly *et al.*, 2023), and low glycemic index (GI) diets (Grden and Jakubczyk, 2023). Legumes have significant advantages due to their positive health benefits and low environmental footprint. They can be cultured in a variety of climates and soil types, demanding less water and fertilizer than other crops, making them a sustainable alternative for farmers. Due to their nutritional and physicochemical properties, legumes are extensively utilized by the food industry. This is attributed to the surging popularity of plant-based diets and the escalating demand for meat alternatives, which presents the opportunity for the development of legume-based meat substitutes (Yanni *et al.*, 2023).

Driving Factors of the GHG Emissions in Crop Production

Fertilization

Numerous empirical studies have robustly demonstrated that fertilization exerts a substantial influence on the carbon footprint (CF) of plant production (Gasol *et al.*, 2012). The application of nitrogen (N) fertilizers in plant cultivation results in the release of gaseous nitrogen compounds. As nitrogen is a chemically reactive compound, it contributes to global warming. The influence of N_2O in contributing to the greenhouse effect is significant due to its high Global Warming Potential (GWP) index of 265. A higher GWP value illustrates that a particular gas possesses a greater warming effect on the earth in comparison to other gases. Ammonia (NH_3), an essential nitrogen-containing compound, exerts a considerable influence on eutrophication and acidification. Substantial emissions of ammonia result from nitrogen-based fertilizers, notably those incorporating urea and ammonium, as well as from animal excrement, posing a grave environmental hazard. To reduce N_2O emanations and counteract water pollution, eutrophication, and acidification caused by gaseous NH_3 emissions, it is important to aim for a closed N cycle on the farm (Emmerling *et al.*, 2020). Taking a comprehensive approach and pinpointing crucial stages in the farm's nitrogen cycle leads to increased nitrogen use efficiency and reduced nitrogen losses in agricultural production (Mahmud *et al.*, 2021).

Fertilization technologies can significantly improve the efficiency of nitrogen use. Key factors in effective fertilization include strip fertilization, timing of application, dose distribution, adapting fertilization levels to variations in soil conditions, and soil nutrient levels (Severin *et al.*, 2016). In the cultivation of canola and wheat, the efficiency of nitrogen use may increase by approximately 10% with the proper fertilization technologies (Sieling and Kage, 2010).

N_2O emissions from nitrogen fertilization can range from 0.77% to 1.25% of the applied nitrogen, for both synthetic and organic fertilizers (Kuikman *et al.*, 2006). As per the findings of the Intergovernmental Panel on Climate Change (IPCC) (Hergoualc'h *et al.*, 2019), the default emission aspect for organic amendments is 1 %, with an uncertainty range spanning from 0.1 % to 1.8 %. When natural fertilizers are used on low-organic matter soils, it can lead to increased N_2O emissions. This happens because fertilizers stimulate microbial activity in the soil by providing carbon compounds, which decrease the oxygen content in the soil and create anaerobic conditions that promote denitrification (Petersen and Sommer, 2011). Additional factors contributing to increased N_2O emissions encompass the timing of natural fertilizer application immediately preceding rainfall, soil pH levels below 6, and the depth of incorporation of natural fertilizers into the soil. Shallow mixing of manure with the soil results in heightened N_2O emissions compared to deep incorporation of the manure (Wang *et al.*, 2021). When slurry is injected into the soil, about 9% of the nitrogen is lost (4.4 % as ammonia (NH_3) and 4.7 % as nitrate (NO_3). However, when the slurry is spread on the field surface, nitrogen losses increase to 27.1 % (20.5 % as ammonia and 6.6 % as nitrate) (Powell *et al.*, 2011).

Greenhouse gas (GHG) emissions resulting from fertilizer production represent the second significant source of emissions in the entire cycle of plant production, following GHG emissions from agricultural fields (Bieńkowski *et al.*, 2014). The amount of emissions produced during the manufacture of mineral fertilizers depends on the effectiveness of the synthesis of N compounds, as well as the usage of natural gas, heat, and electricity. Advancements in fertilizer production technology result in a consistent decrease in these emissions.

Plant Protection

The emission of greenhouse gases (GHGs) in the life cycle of plant protection is primarily linked to the production of plant protection products, fuel combustion from agriculture machinery used during plant protection treatments, and other associated processes. Notably, plant protection contributes less to the formation of carbon footprints (CFs) compared to the fertilization process (Holka *et al.*, 2016; Sow *et al.*, 2023). It is important to note that when plants are effectively protected against pests, they yield more. This means the carbon footprint per product unit (e.g., a kg of wheat grain) is lower. While chemical plant protection products are highly effective, they do pose the risk of dispersing active substances in the environment, potentially contaminating water and soil and accumulating in living organisms (Holka, 2017). On 1 January 2014, the obligation to implement integrated pest management (IPM) principles was introduced in the European Union (EU) (Matyjaszczyk, 2013). As per the IPM plan, it is essential to utilize all available biological, physical, and non-chemical methods before resorting to chemical plant protection. Crop rotation, using appropriate crop varieties, adhering to optimal deadlines, implementing proper agrotechnics, ensuring proper fertilization, and preventing the spread of pests are all crucial components. Crop rotation increases plant species diversity, putting pressure on weed development due to various agronomic factors such as timing of cultivation treatments, soil cultivation systems, intensity of cultivation treatments, diverse active substances, crop residues, plant morphology, and nutrient uptake dynamics. As a result, crop rotation reduces weed infestation, minimizes herbicide use, and enhances crop productivity (Liu *et al.*, 2016). In the experimental framework of cereal cultivation in conjunction with rapeseed, a strategy was implemented involving a 50% reduction in herbicide application, sowing high-yielding cultivars, and an increased seeding density. This method yielded weed control outcomes comparable to those achieved through less intensive monoculture crop cultivation (Harker *et al.*, 2009). Despite integrated pest management (IPM) being recommended, chemical methods remain the primary form of plant protection in conventional agricultural production systems (Piwowar, 2021).

Energy and Machinery Use

The primary form of on-farm energy use is the combustion of fossil fuels. Fossil fuels are utilized in processes related to the production of agricultural machinery, fertilizers, plant protection products, and transportation. The energy consumption in agricultural systems is contingent upon numerous factors, with the most pivotal being soil cultivation, crop rotation, and production intensity. Mechanized cultivation methods have a direct and indirect bearing on greenhouse gas (GHG) emissions (Gołasa *et al.*, 2021).

Direct emissions are primarily determined by the fuel consumption of tractors and self-propelled machinery. The amount of combustion is influenced by factors such as the tractor's power, soil compactness, degree of scaling, depth, and working width of the machines. It is anticipated that fuel demand will rise with deeper cultivation and higher treatment speeds (Moitzi *et al.*, 2014). The average fuel consumption for plowing is approximately 24 liters per hectare (Sarauskis *et al.*, 2012). In Croatia, the conventional tillage method exhibited fuel consumption ranging from 48.1 liters per hectare in the case of barley to 60.99 liters per hectare for maize. The combustion of 1.0 liter of diesel oil was estimated to yield 2.75 kg of CO_2 emissions. Consequently, the CO_2 emissions stemming from fuel consumption under conventional tillage practices varied from 132.36 kg per hectare in barley to 167.72 kg per hectare in maize. Conversely, the utilization of reduced tillage and direct sowing methodologies resulted in fuel conservation and a reduction in CO_2 emissions by 35.3 to 42.9 and 87.8 to 88.1 respectively (Filipovic *et al.*, 2006). In Lithuanian conditions, the cultivation of winter oilseed rape required 23.6 liters of diesel oil per hectare for deep plowing. Significant fuel savings were observed in both direct sowing and strip tilling. In direct sowing, fuel consumption was 3.8 times lower at 6.2 liters per hectare, while in strip tilling and sowing, fuel usage was 3.7 times lower at 6.4 liters per hectare. An assessment of greenhouse gas (GHG) emissions revealed that the use of no-till technologies reduced emissions by 21.2 % compared to technologies based on deep plowing (Saldukaitė *et al.*, 2022).

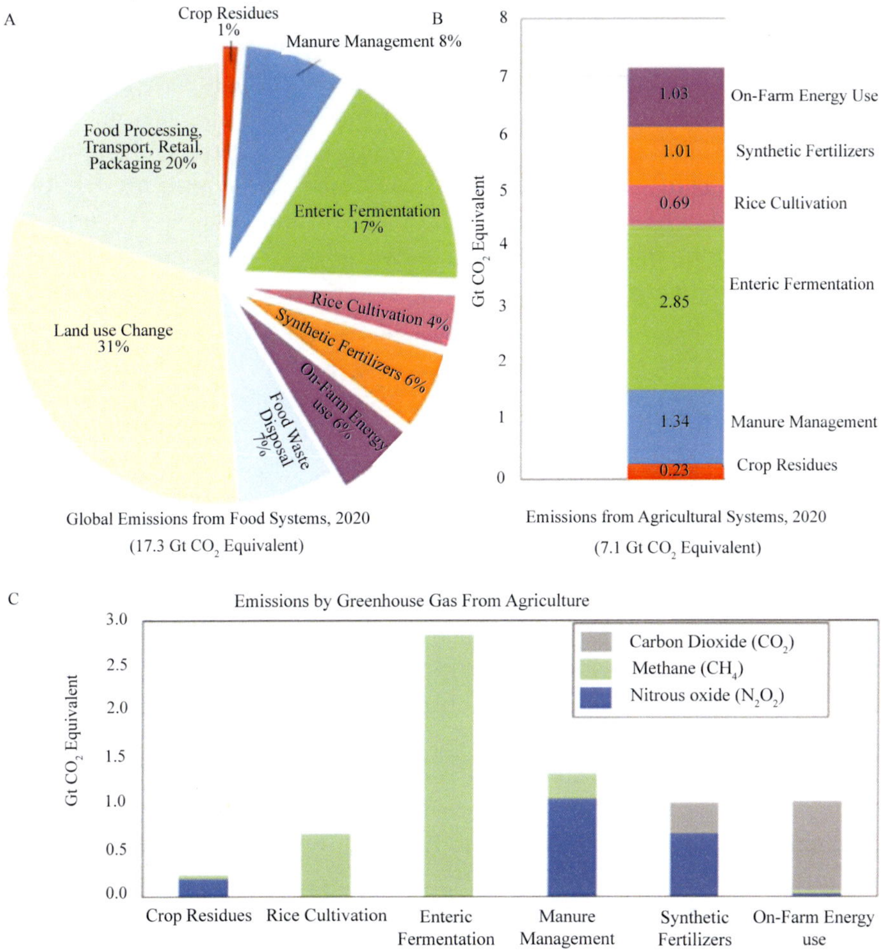

Fig. 12.1: Greenhouse gas emanations from agriculture in 2020. (A) Shares of GHG emissions related to food systems sector-wise (B) Estimates of GHG emanations affiliated with agriculture sector. (C) Breakdown of GHG emissions estimates from agriculture sector (carbon dioxide, methane, nitrous oxide) (Rosa and Gabrielli, 2023).

Legumes as Climate-Smart Crop

Legumes play a pivotal role in advancing food and nutritional security worldwide and possess substantial potential to contribute to global food security and environmental sustainability in the upcoming epoch (Calles *et al.*, 2019). Legumes possess considerable potential and offer opportunities to augment farm income and enhance the risk-bearing capacity of farmers. This can be realized through horizontal crop diversification via the incorporation of pulses into crop rotations. Notably, such practices contribute to the amelioration of human and livestock health by nurturing the soil in situ and mitigating greenhouse gas emissions (Siddique *et al.*, 2012) is increasingly acknowledged. These crops are environmentally sustainable and hold a crucial role in agricultural production. They contribute to the reduction of greenhouse gas emissions by diminishing the need for fertilizer, independently generating nitrogen, and providing residual nitrogen to subsequent crops in the cropping sequence (Mrunalini *et al.*, 2022). The resilience of legumes allows for a significant reduction in production costs, thereby improving food security for the growing population. Legumes can thrive in extreme environments with limited water and have a low carbon footprint, making them a valuable food source for adapting to climate change and variability (Gupta and Kumar, 2019). Legumes use less than half of the non-renewable energy inputs that cereals do. This can help to reduce greenhouse gas emissions. Legumes are becoming increasingly important in climate-resilient agriculture systems due to rising temperatures and limited water resources.

Potential of Legumes to Reduce Greenhouse Gas Emission in Cropping Systems

A significant environmental issue in modern agriculture is the impact of climate change and the increase in greenhouse gas (GHG) concentrations, which has garnered global attention in recent years. Among GHGs, nitrous oxide (N_2O) has gained particular notice, as agricultural soils are the primary source of total global human-caused N_2O emissions, accounting for 75% (FAOSTAT, 2020) and it exhibits higher global warming potential (IPCC, 2014). The increasing global demand for food necessitates higher crop yields, resulting in the extensive utilization of synthetic fertilizer nitrogen, which accounts for 90 % of total fertilizer nitrogen (FAO, 2019). Nonetheless, nitrogen use efficiency varies between 21 and 41 % across different countries worldwide (Omara *et al.*, 2019), indicating that 59-79 % of the nitrogen remains unused in the soil or is released into the environment in the form of NO_x, N_2O, N_2, NO_3, and NH_3. The investigation is currently focused on studying the potential for reducing greenhouse gas (GHG) emissions through the integration of N_2-fixing legumes in cropping systems. The imprints of legume integration in crop rotations in relation to energy-carbon footprints is illustrated in Fig. 2.

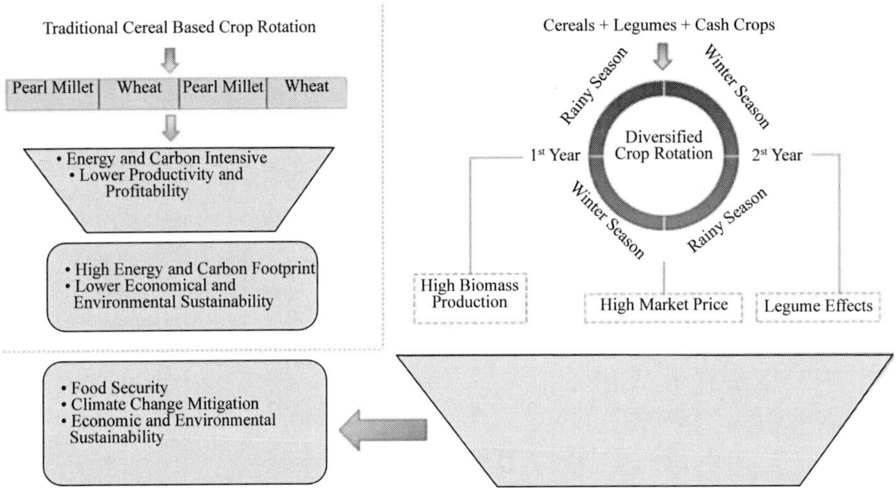

Fig. 12.2: Energy-Carbon Footprint vis-à-vis System Productivity and Profitability of Diversified Crop Rotations (Meena *et al.,* 2022)

Carbon Dioxide Emissions

Nitrogen fixation under a micro-aerobic environment is a highly dynamic process that consumes a significant amount of energy and carbon for every gram of nitrogen assimilation. The process requires 1.77 to 3.01 grams of carbon per gram of nitrogen assimilation associated with nitrogenase activity. Additionally, the respiration of the entire nitrogen-fixing nodules will require between 2.78 and 4.81 grams of carbon per gram of nitrogen (Minchin and Witty, 2005). Globally, the quantity of CO_2 emitted from the production of nitrogen fertilizer via the Haber-Bosch process was equivalent to or less than the CO_2 respired from the nodulated root systems of leguminous plants. Legume plants capture CO_2 through photosynthesis, and the amount of CO_2 released through respiration from their roots is balanced, making it a neutral process (Jensen *et al.*, 2012). However, the CO_2 emitted during the production of fertilizer N is dependent on fossil energy and adds to the overall atmospheric CO_2 levels. Table 1 illustrates the CO_2 emission coefficients in legume production system.

Table 12.1: CO_2 emission coefficients in the legume production system

Inputs	Unit	Emission factor (kg CO_2-eq Unit^{-1})	References
Legumes (seed and output)	kg	0.433	Haq (2014)
Human labor	h	0.7	Houshyar *et al.* (2017)
Machinery	MJ	0.071	Nabavi-Pelesaraei *et al.* (2016)
Diesel fuel	Liter	2.761	Nabavi-Pelesaraei *et al.* (2016)
Fertilizer (N)	kg	1.3	Nabavi-Pelesaraei *et al.* (2016)
Herbicide	MJ	0.069	Audsley *et al.* (2009)

Nitrous Oxide Emissions

Heavy application of nitrogen fertilizer can lead to disturbances in the atmosphere (N_2O emissions), rivers (eutrophication) (Sutton *et al.*, 2011), and human health (methemoglobinemia) (WHO, 2013). Legumes inherently fix atmospheric nitrogen by forming a symbiotic relationship with rhizobium bacteria (Lotjonen and Ollikainen, 2017). This process reduces the nitrogen requirements of both the legume crop and the subsequent crop in a rotational sequence (Autret *et al.*, 2019), making it a potential solution for mitigating N_2O emissions. Based on current understanding, the BNF pathway is no longer regarded as a source of N_2O emissions (Zhong *et al.*, 2009). Nodule decomposition may be mitigated through the use of specific strains of N-fixing bacteria, such as Bradyrhizobium japonicum, in soybean cultivation. This particular bacterium exhibits the capacity to convert N_2O (the terminal phase of denitrification) into N_2 using the N_2O reductase enzyme, facilitated by the presence of nosZ genes (Itakura *et al.*, 2013). This approach has the potential for application to other leguminous crops (Uchida and Akiyama, 2013). The introduction of legumes into the soil triggers immediate mineralization and nitrogen accumulation, attributed to their low carbon-to-nitrogen ratio. The resulting buildup of mineral nitrogen, facilitated by nitrification and denitrification processes, contributes to the loss of N_2O (Jensen *et al.*, 2012). The best approach is to include legumes in crop rotation rather than growing only legumes. This helps reduce the need for mineral nitrogen fertilizers after growing legumes (Reckling *et al.*, 2016). Legumes, when used as part of a crop sequence, help decrease nitrogen runoff and N_2O emissions (Ma *et al.*, 2018; Biernat *et al.*, 2020). Additionally, legumes reduce the ecological impact in a system-based approach through the following ways:

- **Reduced Chemical Fertilizer Requirement:** Nitrogen represents the most widely utilized plant nutrient in the form of synthetic fertilizer and concurrently stands as the most deficient element in soils across the globe. Legumes play a significant role in the nitrogen cycle globally by fixing atmospheric nitrogen through a symbiotic relationship with Rhizobium bacteria in soil ecosystems. This ability to self-sufficiently obtain nitrogen allows legumes to grow in almost any soil without the need for additional fertilizer inputs. From the era of the Green Revolution until 2000, there has been an 80 % increase in nitrogen fertilizer consumption, with half of it being used for major cereal crop production (Canfield *et al.*, 2010). Approximately 40 % of the applied fertilizer to these cereal crops is squandered, which contributes to the generation of N_2O emissions, eutrophication of water bodies, and the formation of hypoxic (low-oxygen) zones (Canfield *et al.*, 2010). Incorporating pulse crops into crop rotation can improve soil organic matter (SOM), enhance water use efficiency, reduce the need for fertilizers, mitigate climate change, boost soil health and system productivity, and break the pest and disease cycle for subsequent crops in rotation (Jangir *et al.*, 2019; Jakhar *et al.*, 2017).

- **Less Water Requirement:** Legumes are a great source of protein and require less water compared to animal-derived protein, according to Alberta Pulse Growers. A study on water footprint revealed that the production of soybeans, chicken, pork, and beef requires 5, 11, 18, and 43 times more water, respectively, compared to legumes. Legume crops had a higher water footprint of 1.59 m3 kg^{-1} compared to cereal crops (1.18 m^3 kg^{-1}) (Ding *et al.*, 2018). The water footprint of various crops has been documented as 200 m^3 ton^{-1} for sugar crops, 300 m^3 ton^{-1} for vegetables, 400 m^3 ton^{-1} for roots and tubers, 1000 m^3 ton^{-1} for fruits, 1600 m^3 ton^{-1} for cereals, 2400 m^3 ton^{-1} for oil crops and 4000 m^3 ton^{-1} for legumes (Mekonnen and Hoekstra, 2011).
- **Carbon Sequestration:** The effects of conservation or no-tillage practices on nitrogen fixation in bean crops have demonstrated that extended use of no-tillage leads to enhanced nodulation. This improvement in nodulation correlates with a notable increase in nitrogen fixation rates over time (Van Kessel and Hartley, 2000). Conventional tillage practices can reduce soil productivity and elevate greenhouse gas emissions due to higher levels of soil disruption and fossil fuel consumption. In contrast, legumes positively affect soil health by increasing soil organic carbon levels, which can lead to better outcomes in subsequent crops within a rotation (Venkatesh *et al.*, 2013). Legumes create favorable conditions for microbial activity, which enhances soil health. Research comparing durum wheat cultivation following legumes to that following cereal crops found a 28% reduction in carbon footprint

(Table 2). The reduction was even more pronounced, approximately 34 %-when durum wheat was grown after a pulse-pulse rotation, compared to a conventional cereal-cereal-durum wheat system (Gan *et al.*, 2012).

Table 12.2: Carbon footprint of cropping systems (Gan *et al.*, 2012)

Cropping system	CO_2 kg^{-1} of grain (kg)	% reduction
Cereal – wheat	0.42	0
Oilseed – wheat	0.34	19
Pulse – wheat	0.30	28

- **Ecological Weed Management:** Legumes possess unique traits such as quick germination, fast growth, early ground coverage, and shading capabilities (Lemerle *et al.*, 2001). These characteristics, along with their high vigor and biomass, contribute to effective weed suppression by establishing competitive niches. Similarly, residues from legume crops can hinder weed germination and growth by creating both physical and chemical barriers (Kocira *et al.*, 2020). Incorporating legumes as intercrops in cropping systems can help reduce weed infestation by competing with weeds. This practice also benefits main crop yields by enhancing nitrogen availability in the soil (Bedoussac *et al.*, 2015).

Main Obstacles to Increase Legume Production and Consumption

Breeding Programs

Over the past 60 years, global food crop production, particularly for cereals has seen remarkable growth despite challenges such as diminishing land availability and rising land costs. During this period, the global population more than doubled, cereal crop production tripled, and the land area used for cultivation increased by only 30 % (Pingali, 2012). Plant breeding programs both before and after the Green Revolution, have predominantly concentrated on cereals, leading to the creation of nitrogen-responsive varieties that increased reliance on chemical fertilizers. This focus has overshadowed the role of legumes and their traditional benefits when paired with cereals. In contrast breeding for legumes has received less investment, and research is ongoing to explore ways to enhance their economic viability (Watson *et al.*, 2017). Significant improvements in yield and stability can be achieved by leveraging the unique traits of each crop to enhance their resistance to biotic and abiotic stresses (Watson *et al.*, 2017). Pea crops often face challenges such as weak standing ability, insufficient ground coverage, limited competitiveness against weeds, and generally low productivity across various soil types (Ferreira *et al.*, 2021). White lupin is noted for its excellent ground coverage, while yellow lupin is recognized for its drought tolerance and suitability for very sandy soils (Watson *et al.*, 2017). Soybean has limited frost tolerance, and common bean's frost tolerance is not well understood. Additionally, addressing specific resistance to insect pests and diseases should be a key focus in future breeding programs (Watson *et al.*, 2017).

Farming Policies and Technologic Barriers

In addition to disparities in breeding efforts, policy, and structural obstacles have limited the responsiveness of crop supply. Policies that favored staple crop production-such as subsidies for fertilizers and credit, price supports, and irrigation infrastructure (especially for rice)-often overlooked traditional non-staple crops like legumes (Welch and Graham, 2000). The development of agricultural practices has largely been driven by the adoption of the most widely used and profitable techniques and crops (Magrini *et al.*, 2018). This trend has resulted in a technological lock-in that favors crop specialization while marginalizing species that are less responsive to cost pressures (Watson *et al.*, 2017). As a result, competition within the agri-food sector led producers to focus on more profitable crops, like cereals, rather than on grain legumes, which typically have lower and more variable yields (Magrini *et al.*, 2016). In this environment, the widespread adoption of agrochemicals became the prevailing approach, often ignoring potential environmental risks (Therond *et al.*, 2017). The absence of technical guidance on using nitrogen-fixing plants such as grain legumes-which can reduce the reliance on synthetic inputs-has significantly influenced these practices (Watson *et al.*, 2017). Research indicates that the low profitability of legumes for farmers can be traced to three main factors: (i) the benefits of legumes are often undervalued because profitability and yield assessments do not account for the benefits of crop rotations; (ii) a lack of interest from the agro-industrial supply chain undermines the perceived value of legumes; and (iii) the low profitability of legumes does not sufficiently offset the reduction in the use of artificial fertilizers (Magrini *et al.*, 2016).

Farmers and businesses face challenges in determining the best market route for legume crops. This process involves identifying the value network structure, establishing collaborative partnerships, and thoroughly understanding the market drivers and barriers specific to their operational context (Hamann *et al.*, 2019). After establishing a market connection, a business must explore ways to sustain and grow its operations, which involves engaging with both upstream and downstream segments of the value network. An effective commercialization strategy for legumes should include plans for scaling up or out, taking into account the capacities of the entire value network (Hamann *et al.*, 2019). For legumes, the absence of economies of scale is frequently cited as a significant barrier to broader adoption. The relatively low production levels of legumes prevent the realization of cost advantages that are typically associated with more productive crops. Increasing legume production can help overcome the economy of scale issue while boosting demand and consumption of locally grown legumes could directly incentivize farmers to grow more of these crops.

Consumers

The adoption of healthier eating habits may be hindered by consumers' limited knowledge of health and nutrition (Magrini *et al.*, 2018). The situation is further complicated by significant variability in general food intake recommendations, especially concerning grain legumes (Marinangeli *et al.*, 2017) which can be extended to cooking skills. Limited knowledge and time constraints, such as long soaking or cooking times, are identified as major obstacles to the regular consumption of legumes (Havemeier *et al.*, 2017). Misconceptions about potential gastrointestinal discomfort from consuming legumes may have caused many consumers to avoid these foods more than necessary (Hall *et al.*, 2017). Finally, the adoption of legumes as alternatives to animal products may be hindered by a general lack of environmental awareness, which is still prevalent in society (Hartmann and Siegrist, 2017). Consumers often underestimate the environmental impact of meat consumption and production, and show a low willingness to alter their meat-eating habits.

Legumes as a Sustainable Alternative

Lower-Carbon Protein Sources

Legumes offer a lower-carbon alternative to animal-based proteins, which contribute to higher carbon emissions through feed production, methane emissions from livestock, and energy use in meat processing. Replacing or supplementing animal proteins with legumes can help lower the overall carbon footprint of food systems (Sow *et al.*, 2024). Legume proteins are increasingly utilized as a viable substitute for animal proteins. For instance, lentil proteins are incorporated into various products such as doughnuts, cakes, and crackers, either to boost protein content or to provide gluten-free options. Additionally, lentils, combined with transglutaminase, are used to create protein-enriched steak or burger substitutes and serve as an emulsifier in salad dressings (Tang *et al.*, 2023). Legumes are also employed as stabilizers in nano-emulsion systems, as ingredients in omega-3 fatty acid supplement capsules, in the creation of antimicrobial membranes, and the manufacture of nanofibers (Khazaei *et al.*, 2019). Legumes contain anti-nutrients like tannins, lectins, phytic acid, and oligosaccharides, which can impact nutrient digestibility and bioavailability. Food processing methods can reduce these anti-nutrients.

Various pre-treatments, such as fermentation with lactic acid bacteria and yeasts, are employed to enhance the nutritional and sensory qualities of legume products, improving their appeal to consumers (Clemente and Jimenez-Lopez,2020).

Biofuels and Renewable Resources

Legumes can be utilized to produce biofuels, providing a more sustainable alternative to fossil fuels. Using legume-based biofuels can decrease dependence on carbon-intensive energy sources and facilitate a shift toward renewable energy. Certain legumes encompassing soybeans, are rich in seed oil and/or starch, making them suitable for biofuel production. Additionally, the legume tree Pongamia, increasingly recognized as a biofuel feedstock, produces seeds with approximately 35-43 % oil, mainly consisting of oleic acid (Biswas *et al.*, 2011). Field pea grain, due to its high starch content, has been explored as a potential feedstock for first-generation bioethanol production, either used alone or in combination with maize grain (Pryor *et al.* 2008). Alfalfa is a promising perennial legume for biofuel and bio-product production due to its potential yield of 4 to 18 metric tons of dry matter per hectare per year, with an average of 8 metric tons per hectare from three to four annual harvests over up to four years (Samac *et al.* 2006). Other perennial legumes, like clovers, are potential feedstocks for second-generation bioethanol, whether cultivated alone or mixed with grasses. Additionally, both herbaceous and woody perennial legumes, with their deep root systems, provide an effective and economical way to remediate excess soil nitrogen, mitigate groundwater nitrate contamination, and reduce the risks associated with rising water tables and dryland salinity (Jensen *et al.*, 2011).

Conclusions

Ensuring food security for a growing global population is increasingly challenging, especially in the context of climate change. Legumes are advantageous crops because they adapt well to diverse environmental conditions and can be incorporated into modern cropping systems to diversify existing agricultural patterns. Creating a climate-smart cropping system is essential. Pulses are well-suited for such systems due to their ability to fix nitrogen biologically, their deep root systems, ground-covering canopies, nutrient-rich biomass, and resilience to stress. Legumes naturally fix atmospheric nitrogen, reducing the need for synthetic fertilizers and thus lowering greenhouse gas emissions. They also enhance soil health and structure, promoting increased carbon sequestration and improved soil fertility, which supports greater agricultural sustainability. The reduced energy use and lower carbon footprint associated with growing legumes are largely due to the diminished need for nitrogen fertilizers, which lowers the nitrogen requirements for subsequent crops. Additionally, the overall energy balance benefits from reduced agrichemical use, as diversifying cropping sequences helps decrease cereal pathogens, pests, and weed populations. With the growing demand for plant-based products such as proteins and oils, and increasing economic and environmental pressures on agro-ecosystems, grain legumes are expected to play a crucial role in future cropping systems. Integrating legumes into low-emission farming practices is a practical approach to reducing carbon footprints and fostering long-term ecological sustainability.

References

Audsley, E., Stacey, K. F., Parsons, D. J., & Williams, A. G. (2009). Estimation of the greenhouse gas emissions from agricultural pesticide manufacture and use.

Autret, B., Beaudoin, N., Rakotovololona, L., Bertrand, M., Grandeau, G., Gréhan, E., et al. (2019). Can alternative cropping systems mitigate nitrogen losses and improve GHG balance? Results from a 19-yr experiment in Northern France. Geoderma, 342, 20-33.

Bedoussac, L., Journet, E. P., Hauggaard-Nielsen, H., Naudin, C., Corre-Hellou, G., Jensen, E. S., et al. (2015). Ecological principles underlying the increase of productivity achieved by cereal-grain legume intercrops in organic farming. A review. Agronomy for Sustainable Development, 35, 911-935.

Bieńkowski, J. F., Dąbrowicz, R., Holka, M., & Jankowiak, J. (2014). The carbon footprint of rapeseed in conventional farming: Case study of large-sized farms in Wielkopolska region (Poland). Asian J. Appl. Sci. Eng, 4, 191-200.

Biernat, L., Taube, F., Loges, R., Kluß, C., & Reinsch, T. (2020). Nitrous oxide emissions and methane uptake from organic and conventionally managed arable crop rotations on farms in Northwest Germany. Sustainability, 12(8), 3240.

Biswas, B., Scott, P. T., & Gresshoff, P. M. (2011). Tree legumes as feedstock for sustainable biofuel production: Opportunities and challenges. Journal of Plant Physiology, 168(16), 1877-1884.

Brentrup, F., Lammel, J., Stephani, T., & Christensen, B. (2018). Updated carbon footprint values for mineral fertilizer from different world regions. In 11th International Conference on Life Cycle Assessment of Food, Kasetsart University (pp. 17-19).

Calles, T., Xipsiti, M., & del Castello, R. (2019). Legacy of the international year of pulses. Environmental Earth Sciences, 78, 1-8.

Canfield, D. E., Glazer, A. N., & Falkowski, P. G. (2010). The evolution and future of Earth's nitrogen cycle. Science, 330(6001), 192-196.

Chai, R., Ye, X., Ma, C., Wang, Q., Tu, R., Zhang, L., & Gao, H. (2019). Greenhouse gas emissions from synthetic nitrogen manufacture and fertilization for main upland crops in China. Carbon balance and Management, 14, 1-10.

Clemente, A., & Jimenez-Lopez, J. C. (2020). Introduction to the Special Issue: Legumes as food ingredient: Characterization, processing, and Applications. Foods, 9(11), 1525.

Ding, D., Zhao, Y., Guo, H., Li, X., Schoenau, J., & Si, B. (2018). Water footprint for pulse, cereal, and oilseed crops in Saskatchewan, Canada. Water, 10(11), 1609.

Emmerling, C., Krein, A., & Junk, J. (2020). Meta-analysis of strategies to reduce NH3 emissions from slurries in European agriculture and consequences for greenhouse gas emissions. Agronomy, 10(11), 1633.

FAO. (2019). World Fertilizer Trends and Outlook to 2022. Rome.

FAOSTAT. (2020). The contribution of agriculture to greenhouse gas emissions. Emission shares.

Ferreira, H., Pinto, E., & Vasconcelos, M. W. (2021). Legumes as a cornerstone of the transition toward more sustainable agri-food systems and diets in Europe. Frontiers in Sustainable Food Systems, 5, 694121.

Filipovic, D., Kosutic, S., Gospodaric, Z., Zimmer, R., & Banaj, D. (2006). The possibilities of fuel savings and the reduction of CO2 emissions in the soil tillage in Croatia. Agriculture, ecosystems & environment, 115(1-4), 290-294.

Gan, Y., Liang, C., Wang, X., & McConkey, B. (2011). Lowering carbon footprint of durum wheat by diversifying cropping systems. Field Crops Research, 122(3), 199-206.

Gao, Y., & Cabrera Serrenho, A. (2023). Greenhouse gas emissions from nitrogen fertilizers could be reduced by up to one-fifth of current levels by 2050 with combined interventions. Nature Food, 4(2), 170-178.

Gasol, C. M., Salvia, J., Serra, J., Antón, A., Sevigne, E., Rieradevall, J., & Gabarrell, X. (2012). A life cycle assessment of biodiesel production from winter rape grown in Southern Europe. Biomass and bioenergy, 40, 71-81.

Gołasa, P., Wysokiński, M., Bieńkowska-Gołasa, W., Gradziuk, P., Golonko, M., Gradziuk, B., et al. (2021). Sources of greenhouse gas emissions in agriculture, with particular emphasis on emissions from energy used. Energies, 14(13), 3784.

Grdeń, P., & Jakubczyk, A. (2023). Health benefits of legume seeds. Journal of the Science of Food and Agriculture, 103(11), 5213-5220.

Grossi, G., Goglio, P., Vitali, A., & Williams, A. G. (2019). Livestock and climate change: impact of livestock on climate and mitigation strategies. Animal Frontiers, 9(1), 69-76.

Gupta, A., & Kumar, S. (2018). Climate resilient agro-technologies for enhanced crop and water productivity under water deficit agro-ecologies Jodhpur: Scientific Publisher, pp. 339-356.

Hall, C., Hillen, C., & Garden Robinson, J. (2017). Composition, nutritional value, and health benefits of pulses. Cereal Chemistry, 94(1), 11-31.

Hamann, K., Vasconcelos, M., Löhrich, N., Odee, D., Vickers, R., & Blazon, N. (2019). A map of value chains for legumes used as food: D4. 1 (D23) for the EU-H2020 project,'. TRansition paths to sUstainable legume-based systems in Europe'(TRUE), funded under Grant Agreement, (727973).

Haq, M. E. (2014). Carbon Footprint of Selected Cereal and Legume Crops Cultivated in the Old Brahmaputra Floodplain Soil. Bangladesh: Bangladesh Agricultural University.

Harker, K. N., O'Donovan, J. T., Irvine, R. B., Turkington, T. K., & Clayton, G. W. (2009). Integrating cropping systems with cultural techniques augments wild oat (Avena fatua) management in barley. Weed Science, 57(3), 326-337.

Hartmann, C., & Siegrist, M. (2017). Consumer perception and behaviour regarding sustainable protein consumption: A systematic review. Trends in Food Science & Technology, 61, 11-25.

Havemeier, S., Erickson, J., & Slavin, J. (2017). Dietary guidance for pulses: The challenge and opportunity to be part of both the vegetable and protein food groups. Annals of the New York Academy of Sciences, 1392(1), 58-66.

Hergoualc'h, K., Akiyama, H., Bernoux, M., Chirinda, N., Prado, A. D., Kasimir, Å., et al. (2019). N_2O emissions from managed soils, and CO_2 emissions from lime and urea application.

Holka, M. (2017). Environmental impact assessment of chemical plant protection in intensive crop production. Journal of Central European Agriculture, 18(3).

Holka, M., Jankowiak, J., Bieńkowski, J., & Dąbrowicz, R. (2016). Life cycle assessment (LCA) of winter wheat in an intensive crop production system in Wielkopolska region (Poland). Appl. Ecol. Environ. Res, 14(535), e545.

Houshyar, E., Mahmoodi-Eshkaftaki, M., & Azadi, H. (2017). Impacts of technological change on energy use efficiency and GHG mitigation of pomegranate: Application of dynamic data envelopment analysis models. Journal of Cleaner Production, 162, 1180-1191

IPCC. (2014). Climate change 2014: Synthesis report. Contribution of Working Groups I, II and III to the Fifth Assessment Report of the Intergovernmental Panel on Climate Change. Core Writing Team. In: Pachauri, R.K., Meyer, L.A. (Eds.), IPCC: Geneva, Switzerland, 151.

Itakura, M., Uchida, Y., Akiyama, H., Hoshino, Y. T., Shimomura, Y., Morimoto, S., et al. (2013). Mitigation of nitrous oxide emissions from soils by Bradyrhizobium japonicum inoculation. Nature Climate Change, 3(3), 208-212.

Jakhar, S. R., Kumar, S., Jangir, C. K., & Meena, R. S. (2017). The role of mycorrhizal relationship in sustainable manner towards plant growth and soil fertility. Indian Journal of Agriculture and Allied Sciences, 3(4), 19-24.

Jangir, C. K., Kumar, S., & Meena, R. S. (2019). Significance of soil organic matter to soil quality and evaluation of sustainability. Sustainable agriculture. Scientific Publisher, Jodhpur, 357-381.

Jensen, E. S., Peoples, M. B., Boddey, R. M., Gresshoff, P. M., Hauggaard-Nielsen, H., JR Alves, B., & Morrison, M. J. (2012). Legumes for mitigation of climate change and the provision of feedstock for biofuels and biorefineries. A review. Agronomy for sustainable development, 32, 329-364.

Kelly, R., Hanus, A., Payne-Foster, P., Calhoun, J., Stout, R., & Sherman, B. W. (2023). Health benefits of a 16-week whole food, high fiber, plant predominant diet among US employees. American Journal of Health Promotion, 37(2), 168-176.

Khazaei, H., Subedi, M., Nickerson, M., Martínez-Villaluenga, C., Frias, J., & Vandenberg, A. (2019). Seed protein of lentils: Current status, progress, and food applications. Foods, 8(9), 391.

Kocira, A., Staniak, M., Tomaszewska, M., Kornas, R., Cymerman, J., Panasiewicz, K., & Lipińska, H. (2020). Legume cover crops as one of the elements of strategic weed management and soil quality improvement. A review. Agriculture, 10(9), 394.

Kuikman, P. J., Van der Hoek, K. W., Smit, A., & Zwart, K. B. (2006). Update of emission factors for nitrous oxide from agricultural soils on the basis of measurements in the Netherlands (No. 1217). Alterra.

Lemerle, D., Verbeek, B., & Orchard, B. (2001). Ranking the ability of wheat varieties to compete with Lolium rigidum. Weed research, 41(3), 197-209.

Liu, C., Cutforth, H., Chai, Q., & Gan, Y. (2016). Farming tactics to reduce the carbon footprint of crop cultivation in semiarid areas. A review. Agronomy for Sustainable Development, 36, 1-16.

Lötjönen, S., & Ollikainen, M. (2017). Does crop rotation with legumes provide an efficient means to reduce nutrient loads and GHG emissions?. Review of agricultural, food and environmental studies, 98, 283-312.

Ma, Y., Schwenke, G., Sun, L., Li Liu, D., Wang, B., & Yang, B. (2018). Modeling the impact of crop rotation with legume on nitrous oxide emissions from rain-fed agricultural systems in Australia under alternative future climate scenarios. Science of The Total Environment, 630, 1544-1552.

Magrini, M. B., Anton, M., Chardigny, J. M., Duc, G., Duru, M., Jeuffroy, M. H., et al. (2018). Pulses for sustainability: breaking agriculture and food sectors out of lock-in. Frontiers in Sustainable Food Systems, 2, 64.

Magrini, M. B., Anton, M., Cholez, C., Corre-Hellou, G., Duc, G., Jeuffroy, M. H., et al. (2016). Why are grain-legumes rarely present in cropping systems despite their environmental and nutritional benefits? Analyzing lock-in in the French agrifood system. Ecological Economics, 126, 152-162.

Mahmud, K., Panday, D., Mergoum, A., & Missaoui, A. (2021). Nitrogen losses and potential mitigation strategies for a sustainable agroecosystem. Sustainability, 13(4), 2400.

Matyjaszczyk, E. (2013). Plant protection in Poland on the eve of obligatory integrated pest management implementation. Pest Management Science, 69(9), 991-995.

Meena, O. P., Sammauria, R., Gupta, A. K., Gupta, K. C., Behera, B., Saxena, R., et al. (2022). Energy-carbon footprint vis-à-vis system productivity and profitability of diversified crop rotations in semi-arid plains of North-West India. Journal of Soil Science and Plant Nutrition, 22(2), 2026-2041.

Mekonnen, M. M., & Hoekstra, A. Y. (2011). The green, blue and grey water footprint of crops and derived crop products. Hydrology and earth system sciences, 15(5), 1577-1600.

Minchin, F. R., & Witty, J. F. (2005). Plant respiration, 195-205.

Mirzabaev, A., Olsson, L., Kerr, R. B., Pradhan, P., Ferre, M. G. R., & Lotze-Campen, H. (2023). Climate change and food systems. Science and innovations for food systems transformation, 511.

Moitzi, G., Wagentristl, H., Refenner, K., Weingartmann, H., Piringer, G., Boxberger, J., & Gronauer, A. (2014). Effects of working depth and wheel slip on fuel consumption of selected tillage implements. Agricultural Engineering International: CIGR Journal, 16(1), 182-190.

Mrunalini, K., Behera, B., Chandana, P., Patnaik, G. P., Modi, R. U., Saraswat, A., et al. (2022). Legumes to reduce ecological footprints for climate-smart cropping systems. In Advances in Legumes for Sustainable Intensification (pp. 403-420). Academic Press.

Nabavi-Pelesaraei, A., Hosseinzadeh-Bandbafha, H., Qasemi-Kordkheili, P., Kouchaki-Penchah, H., & Riahi-Dorcheh, F. (2016). Applying optimization techniques to improve of energy efficiency and GHG (greenhouse gas) emissions of wheat production. Energy, 103, 672-678.

Omara, P., Aula, L., Oyebiyi, F., & Raun, W. R. (2019). World cereal nitrogen use efficiency trends: review and current knowledge. Agrosystems, Geosciences & Environment, 2(1), 1-8.

Özköse, A. (2024). Contribution of Legume-Derived Biological Nitrogen Fixation in Reducing Greenhouse Gas Emissions Originating from Agriculture in Türkiye. Polish Journal of Environmental Studies, 33(2).

Pachauri, R.K., Meyer, L.A. (Eds.), IPCC: Geneva, Switzerland, 151.

Petersen, S. O., & Sommer, S. G. (2011). Ammonia and nitrous oxide interactions: roles of manure organic matter management. Animal Feed Science and Technology, 166, 503-513.

Pingali, P. L. (2012). Green revolution: impacts, limits, and the path ahead. Proceedings of the national academy of sciences, 109(31), 12302-12308.

Piwowar, A. (2021). The use of pesticides in Polish agriculture after integrated pest management (IPM) implementation. Environmental Science and Pollution Research, 28(21), 26628-26642.

Poore, J., & Nemecek, T. (2018). Reducing food's environmental impacts through producers and consumers. Science, 360(6392), 987-992.

Powell, J. M., Jokela, W. E., & Misselbrook, T. H. (2011). Dairy slurry application method impacts ammonia emission and nitrate leaching in no-till corn silage. Journal of environmental quality, 40(2), 383-392.

Pryor, S. W., Lenling, M., & Wiesenborn, D. P. (2008). Integrated use of field pea starch and corn for ethanol production. In 2008 Providence, Rhode Island, June 29–July 2, 2008 (p. 1). American Society of Agricultural and Biological Engineers.

Reckling, M., Bergkvist, G., Watson, C. A., Stoddard, F. L., Zander, P. M., Walker, R. L., et al. (2016). Trade-offs between economic and environmental impacts of introducing legumes into cropping systems. Frontiers in Plant Science, 7, 669.

Rees, K., Takeda, A., Martin, N., Ellis, L., Wijesekara, D., Vepa, A., et al. (2019). Mediterranean□ style diet for the primary and secondary prevention of cardiovascular disease. Cochrane Database of Systematic Reviews, 3.

Rosa, L., & Gabrielli, P. (2023). Achieving net-zero emissions in agriculture: a review. Environmental Research Letters, 18(6), 063002.

Saldukaitė, L., Šarauskis, E., Zabrodskyi, A., Adamavičienė, A., Buragienė, S., Kriaučiūnienė, Z., & Savickas, D. (2022). Assessment of energy saving and GHG reduction of winter oilseed rape production using sustainable strip tillage and direct sowing in three tillage technologies. Sustainable Energy Technologies and Assessments, 51, 101911.

Samac, D. A., Jung, H. J. G., & Lamb, J. F. (2016). Development of alfalfa (Medicago sativa L.) as a feedstock for production of ethanol and other bioproducts. In Alcoholic fuels (pp. 79-98). CRC Press.

Samanta, S., Kumar, S., Khedwal, R. S., & Jakhar, S. R. (2017). Allelopathy for sustainable weed management. Journal of Pharmacognosy and Phytochemistry, 832-837.

Sarauskis, E., Buragiene, S., Romaneckas, K., Sakalauskas, A., Jasinskas, A., Vaiciukevicius, E., & Karayel, D. (2012). Working time, fuel consumption and economic analysis of different tillage and sowing systems in Lithuania. Eng. Rural Dev, 11, 52-59.

Severin, M., Fuss, R., Well, R., Garlipp, F., & Van den Weghe, H. (2016). Soil, slurry and application effects on greenhouse gas emissions.

Siddique, K. H., Johansen, C., Turner, N. C., Jeuffroy, M. H., Hashem, A., Sakar, D., et al. (2012). Innovations in agronomy for food legumes. A review. Agronomy for sustainable development, 32, 45-64.

Sieling, K., & Kage, H. (2010). Efficient N management using winter oilseed rape. A review. Agronomy for Sustainable Development, 30, 271-279.

Sow, S., Ranjan, S., Behera, B., Ghosh, M., Kumar, S., & Dutta, S. K. (2023). Maintaining agricultural sustainability through carbon footprint management. Current Science, 125(9), 939-944.

Sow, S., Ranjan, S., Padhan, S.R., Nath, D., & Kumar, N. (2024). Agroforestry and Soil Carbon Sequestration: A Nexus for System Sustainability. In: Kumar, S., Alam, B., Taria, S., Singh, P., Yadav, A., Arunachalam, A. (eds) Agroforestry Solutions for Climate Change and Environmental Restoration. Springer, Singapore, pp 103-126.

Sutton, M. A., Howard, C. M., Erisman, J. W., Billen, G., Bleeker, A., Grennfelt, P., et al. (Eds.). (2011). The European nitrogen assessment: sources, effects and policy perspectives. Cambridge university press.

Tang, Q., Roos, Y. H., & Miao, S. (2023). Effect of microbial transglutaminase on functional, rheological, and structural properties of lentil protein-casein binary gels. Food Hydrocolloids, 143, 108838.

Theodoridis, X., Triantafyllou, A., Chrysoula, L., Mermigkas, F., Chroni, V., Dipla, K., et al. (2023). Impact of the level of adherence to the DASH diet on blood pressure: a systematic review and meta-analysis. Metabolites, 13(8), 924.

Therond, O., Duru, M., Roger-Estrade, J., & Richard, G. (2017). A new analytical framework of farming system and agriculture model diversities. A review. Agronomy for sustainable development, 37, 1-24.

Tilman, D., & Clark, M. (2014). Global diets link environmental sustainability and human health. Nature, 515(7528), 518-522.

Tubiello, F. N., Rosenzweig, C., Conchedda, G., Karl, K., Gütschow, J., Xueyao, P., et al. (2021). Greenhouse gas emissions from food systems: building the evidence base. Environmental Research Letters, 16(6), 065007.

Uchida, Y., & Akiyama, H. (2013). Mitigation of postharvest nitrous oxide emissions from soybean ecosystems: a review.

Van Kessel, C., & Hartley, C. (2000). Agricultural management of grain legumes: has it led to an increase in nitrogen fixation?. Field Crops Research, 65(2-3), 165-181.

Venkatesh, M. S., Hazra, K. K., Ghosh, P. K., Praharaj, C. S., & Kumar, N. (2013). Long-term effect of pulses and nutrient management on soil carbon sequestration in Indo-Gangetic plains of India. Canadian Journal of Soil Science, 93(1), 127-136.

Wang, C., Amon, B., Schulz, K., & Mehdi, B. (2021). Factors that influence nitrous oxide emissions from agricultural soils as well as their representation in simulation models: a review. Agronomy, 11(4), 770.

Wang, J., & Zhang, Y. X. (2023). A Study on Influence of Land Morphology on Carbon Emissions: The Case of Yangtze River Delta Region. Polish Journal of Environmental Studies, 32(6).

Watson, C. A., Reckling, M., Preissel, S., Bachinger, J., Bergkvist, G., Kuhlman, T., et al. (2017). Grain legume production and use in European agricultural systems. Advances in Agronomy, 144, 235-303.

Welch, R. M., & Graham, R. D. (2000). A new paradigm for world agriculture: productive, sustainable, nutritious, healthful food systems. Food and Nutrition Bulletin, 21(4), 361-366.

WHO. (2013). Review of evidence on health aspects of air pollution – RE-VIHAAP Project: Technical Report. WHO Regional Office for Europe, Copenhagen.

Willett, W., Rockström, J., Loken, B., Springmann, M., Lang, T., Vermeulen, S., et al. (2019). Food in the Anthropocene: the EAT–Lancet Commission on healthy diets from sustainable food systems. The lancet, 393(10170), 447-492.

Yadav, S. S., Hunter, D., Redden, B., Nang, M., Yadava, D. K., & Habibi, A. B. (2015). Impact of climate change on agriculture production, food, and nutritional security. Crop wild relatives and climate change, 1-23.

Yanni, A. E., Iakovidi, S., Vasilikopoulou, E., & Karathanos, V. T. (2023). Legumes: A Vehicle for Transition to Sustainability. Nutrients, 16(1), 98.

Zhong, Z., Lemke, R. L., & Nelson, L. M. (2009). Nitrous oxide emissions associated with nitrogen fixation by grain legumes. Soil Biology and Biochemistry, 41(11), 2283-2291.

13

Application of Remote Sensing and Image Analysis in Plant Protection for Grain Legume Based Cropping System

Jaya Sinha[1*] and Lalita Rana[2]

[1]*College of Agricultural Engineering and Technology, Dr. Rajendra Prasad Central Agricultural University, Pusa 848125, Bihar*

[2]*Sugarcane Research Institute, Dr. Rajendra Prasad Central Agricultural University Pusa 848125, Bihar*

**Corresponding author's email: jaya.sinha@rpcau.ac.in*

Abstract

Remote sensing and image analysis have emerged as essential tools in modern agriculture, particularly for plant protection in grain-based legume cropping systems. These technologies are increasingly valuable in addressing challenges such as climate change, pest infestations, and the demand for sustainable practices. Grain legumes, including peas, beans, lentils, and chickpeas, are vital for global food security and soil health due to their protein content and nitrogen-fixing capabilities. However, their productivity is often threatened by pests, diseases, and weeds. Remote sensing allows for early detection of crop stress through analysis of specific wavelengths of light, enabling farmers to identify issues before visible symptoms appear. Tools like the normalized difference vegetation index (NDVI) provide insights into crop health, allowing for timely and targeted interventions. Hyperspectral and multispectral imaging further enhance disease detection by identifying physiological changes in crops that are not visible to the naked eye. In addition to disease monitoring, remote sensing aids in pest management by detecting changes in plant temperature profiles and utilizing drone surveillance for real-time data collection. Image analysis complements these efforts by offering detailed insights into crop conditions, enabling precision agriculture practices that minimize environmental impact and reduce costs. By integrating remote sensing and image analysis with other technologies like GIS and IoT, farmers can achieve more efficient, sustainable farming practices, ultimately leading to higher yields and better protection for grain legume crops.

Keywords: *Food security, Remote sensing, GIS, Grain legumes, IoT, Pest management*

Introduction

Remote sensing and image analysis have become invaluable tools in modern agriculture, particularly in plant protection for grain-based legumes cropping systems. As agriculture continues to face challenges related to climate change, pest infestations, and the need for sustainable practices, the integration of these advanced technologies offers significant benefits. Grain-based legumes, such as peas, beans, lentils, and chickpeas, are critical components of global agriculture (Foyers *et al.*, 2016). These crops are essential for food security providing protein-rich food sources and playing a crucial role in sustainable farming through nitrogen fixation, which enriches the soil for subsequent crops. However, the productivity of these legume crops is often threatened by various biotic stresses, including pests, diseases, and weeds. Effective plant protection strategies are therefore necessary to ensure the health and yield of these crops. In recent years, the application of remote sensing, integrated pest management and image analysis technologies has revolutionized plant protection practices, offering new, efficient, and precise tools for monitoring and managing threats in grain-based legume cropping systems.

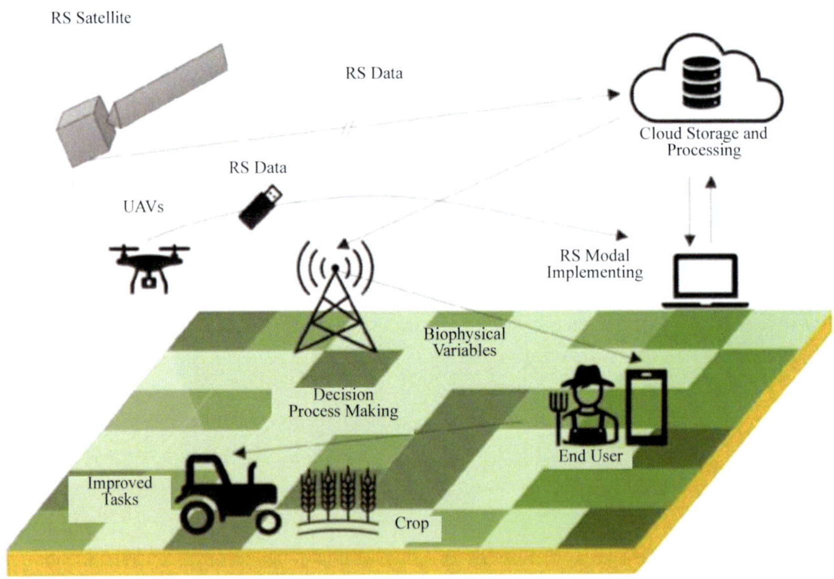

Fig. 13.1: Application of remote sensing in agriculture

Fig. 13.2: Application of drone in agriculture

Challenges in Legumes Production

The primary challenges in grain-based legumes production include:

Pests and Diseases: Legumes are prone to various pests such as aphids, pod borers, and root-knot nematodes, as well as diseases like blight, rust, and mildew.

Climate Variability: Changes in temperature, precipitation patterns, and extreme weather events can stress legumes, making them more vulnerable to pests and diseases.

Soil Degradation: Continuous cropping of legumes can lead to soil degradation, affecting crop health and yield.

Remote Sensing in Agriculture

Remote sensing refers to the acquisition of information about an object or phenomenon without making physical contact with it. In agriculture, remote sensing is used to monitor crop health, assess soil conditions, and predict yields by collecting data from satellites, drones, or other aerial platforms. The data collected through remote sensing can be processed and analyzed to make informed decisions about plant protection, early disease detection (Sow and Ranjan, 2021).

Importance in Plant Protection

Plant protection involves strategies and practices aimed at preventing or mitigating damage to crops caused by pests, diseases, and environmental stressors. Remote sensing technologies provide a non-invasive method to detect early signs of stress in crops, allowing for timely interventions that can save yields and reduce the need for chemical inputs. For grain-based legumes, remote sensing provides a means to monitor vast areas of farmland continuously and in real time, facilitating early intervention and reducing the spread of diseases or pest infestations.

One of the primary benefits of remote sensing in plant protection is its ability to detect changes in crop physiology before visible symptoms appear. This is made possible by the analysis of specific wavelengths of light reflected by the crops, particularly in the infrared and near-infrared spectra. Healthy plants reflect light differently than stressed plants, allowing for the early detection of issues such as fungal infections, insect damage, or nutrient deficiencies. For example, normalized difference vegetation index (NDVI) is a common metric derived from remote sensing data that can indicate the overall health and vigor of crops. By regularly monitoring NDVI values, farmers can identify areas of concern and take targeted action to mitigate damage (Mulla, 2013). Remote sensing can detect early signs of disease in legume crops by identifying changes in plant physiology that are not visible to the naked eye. Hyperspectral and multispectral imaging are particularly useful in this regard.

Hyper spectral Imaging

This technology captures a wide range of wavelengths beyond the visible spectrum, allowing for the detection of specific disease markers. For example, hyper spectral imaging can identify changes in leaf pigmentation caused by fungal infections before symptoms are visually apparent.

Multispectral Imaging

By capturing images in specific bands of the electromagnetic spectrum, multispectral imaging can help in detecting stress responses in plants, such as chlorophyll degradation or water stress, which are indicative of disease.

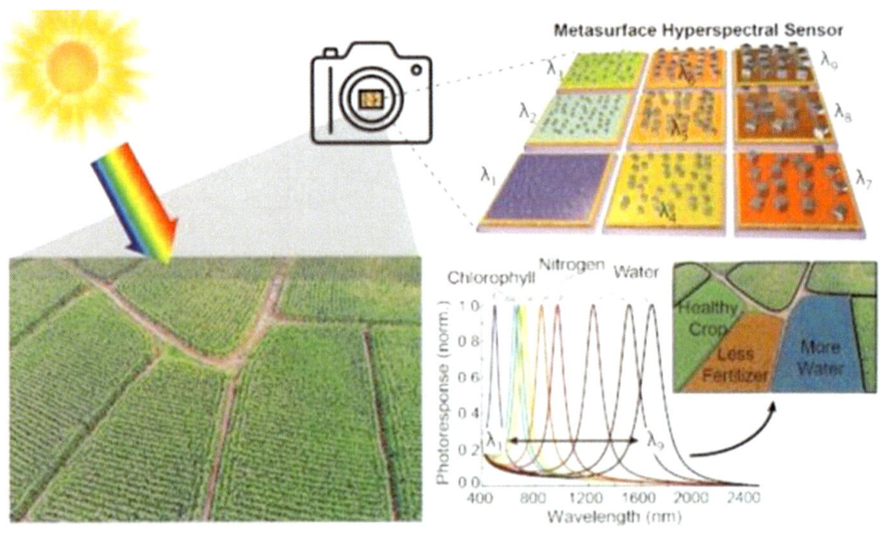

Fig. 13.3: Hyper spectral imaging

Fig. 13.4: Multispectral imaging

A study on chickpea crops showed that remote sensing could detect Ascochyta blight several days before symptoms were visible, allowing for early intervention and reducing yield loss.

Table 13.1: Different types of hyperspectral and multispectral sensors used in agriculture

Sensor Type	Sensor Name	Wavelength Range (nm)	Spectral Bands	Applications in Agriculture
Hyperspectral	AVIRIS	400 - 2500	224	Crop health monitoring, soil property analysis, disease detection
Hyperspectral	Hyperion	400 - 2500	220	Vegetation analysis, water content measurement, nutrient management
Hyperspectral	HyMap	400 - 2500	126	Crop yield estimation, stress detection, soil property mapping
Multispectral	Landsat 8 OLI	430 - 1250	9	Land cover classification, crop monitoring, vegetation indices calculation
Multispectral	Sentinel-2 MSI	443 - 2190	13	Crop monitoring, soil health assessment, precision farming
Multispectral	RapidEye	440 - 850	5	Crop classification, stress detection, biomass estimation
Multispectral	Planet Scope	455 - 860	4	Field-level monitoring, vegetation health assessment, crop growth analysis
Multispectral	World View-3	400 - 1040	16	High-resolution crop mapping, stress detection, water management
Multispectral	UAV-based Multispectral Sensors (e.g., MicaSense RedEdge)	475 - 840	5	Precision agriculture, field variability assessment, crop disease detection

Pest Monitoring and Management

Remote sensing aids in the monitoring of pest populations and the assessment of crop damage caused by pests. This technology can be integrated with other tools, such as pheromone traps, to enhance pest management strategies.

Thermal Imaging: Pests often cause changes in the thermal profile of plants. Thermal imaging can detect these changes, enabling the identification of pest-infested areas within a field. This allows for targeted pest control measures, reducing the use of pesticides and minimizing environmental impact.

NDVI (Normalized Difference Vegetation Index): NDVI is a commonly used remote sensing index that measures plant vigor by comparing the reflectance of red and near-infrared light. A decline in NDVI values can indicate pest damage, enabling farmers to take action before the damage becomes widespread.

Drone Surveillance: Drones equipped with remote sensing cameras can provide real-time data on pest infestations over large areas. This rapid assessment capability is especially useful in large-scale legume production, where ground-based monitoring would be time-consuming and less effective.

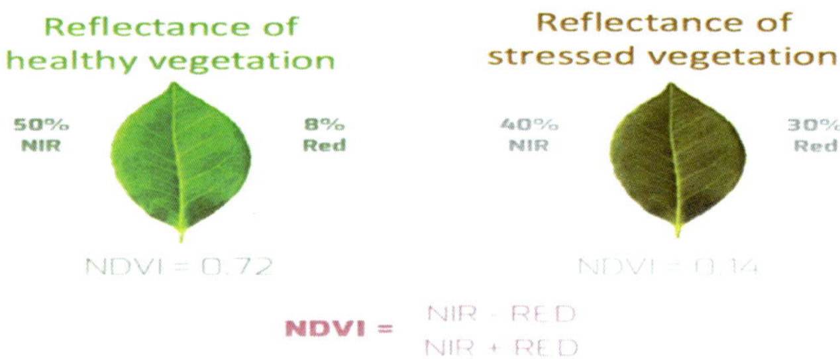

Fig. 13.5: NDVI (Normalized Difference Vegetation Index)

Fig. 13.6: Drone Surveillance

Soil Health and Nutrient Management

Remote sensing can also be used to assess soil health, which is crucial for the successful cultivation of legumes. Healthy soil supports robust plant growth and helps in resisting pests and diseases.

Soil Moisture Monitoring: Remote sensing technologies, such as microwave radiometry, can monitor soil moisture levels across large fields. Maintaining optimal soil moisture is important for legumes, as water stress can make them more susceptible to diseases like root rot.

Nutrient Deficiency Detection: Multispectral and hyperspectral imaging can detect nutrient deficiencies in legumes by analyzing changes in leaf color and texture. For instance, nitrogen deficiency in legumes can be detected by a reduction in chlorophyll content, which appears as a specific signature in remote sensing data (Ranjan *et al.*, 2022).

Image Analysis in Plant Protection for Legumes

Image analysis complements remote sensing by providing more detailed information about the condition of crops at the ground level. Advanced image processing techniques, including machine learning and computer vision, can analyze high-resolution images captured by drones or ground-based cameras to identify specific types of pests or diseases affecting grain-based legumes. This level of detail allows for precision agriculture practices, where interventions such as pesticide application can be targeted specifically to affected areas rather than treating the entire field. This not only reduces costs and labor but also minimizes the environmental impact of agricultural practices.

For instance, image analysis algorithms can be trained to recognize symptoms of common diseases in grain-based legumes, such as Ascochyta blight in chickpeas or Anthracnose in beans, based on patterns and color changes in the leaves. Similarly, these algorithms can identify insect damage by detecting characteristic feeding patterns or discoloration. As these technologies evolve, they are becoming increasingly capable of distinguishing between different stress factors, allowing for more accurate diagnosis and tailored treatment strategies (Sankaran *et al.*, 2010).

Automated Disease Detection

Image analysis algorithms can be trained to recognize specific disease symptoms in legume crops. These algorithms use machine learning models that have been trained on large datasets of diseased and healthy plants.

Convolutional Neural Networks (CNNs): CNNs are a type of deep learning algorithm that excel in image recognition tasks. In agriculture, CNNs can be used to automatically detect diseases in legume crops by analyzing patterns in leaf images. This automation reduces the need for manual scouting and speeds up the disease management process (Sow *et al.*, 2024).

Precision Agriculture Tools: By integrating image analysis with precision agriculture tools, farmers can receive real-time alerts about disease outbreaks in specific areas of their fields. This allows for localized treatment, which is more effective and less costly than blanket applications of fungicides.

Pest Identification and Damage Assessment

Image analysis is also crucial in identifying pests and assessing the extent of damage they cause to legume crops.

Object Detection Algorithms: These algorithms can identify specific pests on legume plants by analyzing images captured by drones or ground-based cameras. For instance, an algorithm can be trained to recognize aphids on bean plants by their shape and color (Weis *et al.*, 2017).

Damage Quantification: Image analysis can quantify the damage caused by pests by comparing images of healthy plants with those that have been affected. This quantification helps in assessing the economic impact of pest infestations and in planning effective control measures.

Monitoring Crop Growth and Health

Image analysis tools are essential for monitoring the overall growth and health of legume crops throughout the growing season.

Vegetation Indices Calculation: Various vegetation indices, such as NDVI and SAVI (Soil-Adjusted Vegetation Index), can be calculated from remote sensing images to assess plant health. Image analysis software processes these indices to provide insights into plant vigor, biomass, and potential yield.

Growth Stage Identification: By analyzing time-series images, image analysis tools can track the growth stages of legume crops. This information is important for scheduling interventions such as irrigation, fertilization, and pest control at the optimal time.

Integration of Remote Sensing and Image Analysis with Other Technologies

The full potential of remote sensing and image analysis in plant protection for legumes is realized when these technologies are integrated with other agricultural innovations.

Geographic Information Systems (GIS)

GIS can be used to map and analyze spatial data related to legume crops. By integrating remote sensing data with GIS, farmers can visualize the spatial distribution of diseases, pests, and soil conditions across their fields.

Spatial Analysis: GIS tools can perform spatial analysis to identify areas within a field that are more prone to disease or pest infestations. This analysis helps in optimizing the placement of sensors and the application of control measures.

Variable Rate Technology (VRT): By combining GIS with remote sensing data, farmers can implement VRT, which allows for the precise application of inputs like fertilizers and pesticides based on the specific needs of different areas within a field.

Internet of Things (IoT)

The Internet of Things (IoT) connects various agricultural devices and sensors, enabling real-time monitoring and data collection.

Sensor Networks: IoT sensors placed in fields can continuously monitor environmental conditions, such as temperature, humidity, and soil moisture. When combined with remote sensing data, these sensors provide a comprehensive view of the factors affecting legume crop health.

Automated Decision-Making: IoT systems can be programmed to automatically trigger interventions based on the data received from remote sensing and image analysis tools. For example, an IoT system could activate an irrigation system when soil moisture levels fall below a certain threshold (Wolfert *et al.*, 2017).

Benefits of Remote Sensing and Image Analysis in Plant Protection for Legumes

The use of remote sensing and image analysis in plant protection offers numerous benefits, particularly in the context of sustainable agriculture.

- **Early Detection and Prevention:** One of the most significant advantages of these technologies is the ability to detect plant stress and disease at an early stage. Early detection allows for timely interventions, which can prevent the spread of diseases and reduce crop losses.
- **Precision in Pest and Disease Management:** Remote sensing and image analysis enable precision agriculture, where interventions are tailored to the specific needs of different areas within a field. This precision reduces the use of chemical inputs, lowering production costs and minimizing environmental impact.
- **Enhanced Decision-Making:** The data generated by remote sensing and image analysis tools provides farmers with actionable insights that improve decision-making. These technologies enable farmers to make informed choices about when and where to apply fertilizers, pesticides, and other inputs, leading to more efficient and sustainable farming.

Future Prospects

The future prospects of remote sensing and image analysis in agriculture particularly for grain-based legume cropping systems, are promising as technological advancements continue to evolve. As remote sensing technologies become more sophisticated, the precision of crop monitoring and management will improve. This will enable farmers to make even more accurate and timely decisions, reducing the reliance on chemical inputs and enhancing sustainability. The integration of AI with remote sensing and image analysis will lead to more advanced predictive models for crop health and yield forecasting. AI algorithms can process vast amounts of data from remote sensing platforms to identify patterns and trends, providing farmers with actionable insights in real-time.

As the cost of remote sensing technologies decreases, these tools will become more accessible to smallholder farmers. Scalable solutions tailored to smaller plots of land will help ensure that even farmers with limited resources can benefit from precision agriculture. Remote sensing will play a critical role in helping farmers adapt to climate change by providing detailed information on soil moisture, temperature, and other environmental factors. This data will be crucial for developing resilient cropping strategies that can withstand climate variability. As these technologies continue to advance, their contribution to improving the productivity and resilience of grain-based legume cropping systems will be vital in addressing global food security challenges. With a growing global population and increasing environmental pressures, remote sensing and image analysis will be key in ensuring sustainable agricultural practices.

Conclusion

The application of remote sensing and image analysis in plant protection is transforming the way grain-based legume cropping systems are managed. These technologies provide farmers with powerful tools to monitor crop health, detect early signs of stress, and implement targeted interventions. By enhancing the precision and efficiency of plant protection practices, remote sensing, and image analysis contribute to higher yields, reduced environmental impact, and more sustainable agricultural practices. As these technologies continue to advance, their role in safeguarding the future of grain-based legume production will only become more critical.

References

Foyer, C. H., Lam, H. M., Nguyen, H. T., Siddique, K. H., Varshney, R. K., Colmer, T. D., et al. (2016). Neglecting legumes has compromised human health and sustainable food production. Nature Plants, 2(8), 1-10.

Mulla, D. J. (2013). Twenty five years of remote sensing in precision agriculture: Key advances and remaining knowledge gaps. Biosystems Engineering, 114(4), 358-371

Ranjan, S., Sow, S., Kumari, J., Kumar, N., Roy, D.K., & Pramanick, B. (2022). Use of Remote Sensing in Agriculture. Agriculture Letters, 3(9), 48-52.

Sankaran, S., Mishra, A., Ehsani, R., & Davis, C. (2010). A review of advanced techniques for detecting plant diseases. Computers and Electronics in Agriculture, 72(1), 1-13. doi:10.1016/j.compag.2010.02.007.

Sow, S., & Ranjan, S. (2021). Popularization of Pulse Production for Ensuring Nutritional and Food Security. Agriculture & Food E:Newsletter, 3(2), 16-19.

Sow, S., Ranjan, S., Seleiman, M.F., Alkharabsheh, H.M., Kumar, M., et al. (2024). Artificial intelligence for maximizing agricultural input use efficiency: exploring nutrient, water and weed management strategies. Phyton-International Journal of Experimental Botany, 93(7), 1569-1598. doi:10.32604/phyton.2024.052241.

Weis, M., Rumpf, T., Plümer, L., & Sökefeld, M. (2017). Improving image-based detection of crop diseases by understanding image acquisition. Plant Methods, 13(1), 1-10. doi:10.1186/s13007-017-0254-7

14

Role of Grain Legumes to Achieve Food and Nutritional Security

Rina Roy[1*], Manish Kumar[2], Sweeti Kumari[3], Shweta Kumari[2] and Devendra Mandal[1]

[1]Department of Agronomy, Dr. Kalam Agricultural College, Kishanganj Bihar Agricultural University, Sabour, Bihar, India

[2]Department of Soil Science and Agricultural Chemistry, Dr. Kalam Agricultural College, Kishanganj, Bihar Agricultural University, Sabour, Bihar, India

[3]Department of Agronomy, Veer Kunwar Singh College of Agriculture, Dumraon Buxar, Bihar, India

**Corresponding author's email: r2.jnkvv@gmail.com*

Abstract

The United Nations predicts that by 2050, the global population will reach 9.6 billion, leading to urgent challenges such as ensuring food security and reducing greenhouse gas emissions. Climate change, coupled with biotic and abiotic pressures, will significantly impact global food production. Despite advances in agriculture, malnutrition remains a widespread issue, affecting millions, especially in developing countries. Protein-energy malnutrition, which affects over 170 million children and women, poses a significant threat to national economic development, increasing healthcare costs and reducing productivity. One key solution to combating malnutrition lies in promoting legumes (pulses), which are nutrient-dense, protein-rich, and vital for both human and animal consumption. Pulses have been a staple in diets for centuries, providing essential nutrients like protein, fiber, vitamins, and minerals. They are not only beneficial for human health, reducing the risk of heart disease, diabetes, and cancer, but they also contribute to sustainable agriculture. Legumes can fix atmospheric nitrogen, reducing the need for synthetic fertilizers, improving soil health, and increasing agricultural productivity. However, despite their benefits, pulse production lags behind other staple crops like cereals. With proper management, legumes could play a vital role in alleviating global hunger, particularly in developing countries. Increased awareness and efforts to incorporate legumes into farming systems and daily diets could help meet the growing global demand for nutritious, sustainable food.

Keywords: *Climate change, Food security, Global hunger, Legumes, Protein-energy malnutrition, Sustainable agriculture*

Introduction

The United Nations (2013) estimates that world's population will be 9.6 billion by 2050 presenting several critical global challenges in which most urgent challenges are ensuring food security and reducing net greenhouse gas emissions to mitigate climate change. Global food production will be significantly affected by Climate change along with increasing biotic and abiotic pressures on crop systems. As we know that healthy diet is essential for human health and national development since healthy people are more productive workers who can earn higher wages and improve their livelihood. Despite higher agricultural production, billions of people still affected by malnutrition in our world (Webb *et al.,* 2018). Malnutrition found in every country in one or another form, regardless of its development status but developing nations are more susceptible to it. Protein-energy malnutrition (PEM) significantly impacts over 170 million preschool children and lactating women in developing countries across Africa and Asia (Khalid and Elharadallou, 2013; Nedumaran *et al.,* 2015). The economics of the country as well as the growth and development of children are seriously threatened by malnutrition. Malnutrition increases the risk of illness and death in children and diminishes their learning ability (WHO, 2010). A child suffering from malnutrition often becomes an adult with poor work performance. Prolonged malnourishment causes weakened immune systems and recurrent infections, which increase medical expenses and eventually hinder the country's economic development (Blössner and de Onis, 2005). Therefore, to address these challenges, it is essential to promote a balanced and sustainable diversification of consumption patterns in which legumes play an important role due to their nutritional and health benefits. Thus, one of the best ways to reduce poverty and alleviating malnutrition is to combine pulses with cereals in their diet.

Grain legumes, commonly known as pulses, play a pivotal role in global agriculture due to their multifaceted benefits in both food production and nutritional value. They are known for their protein-rich composition, capability to fix atmospheric nitrogen, and adaptability to marginal environments (Sow and Ranjan, 2021). In the face of escalating global food insecurity and malnutrition challenges, grain legumes offer a strategic solution for improving food availability and dietary quality, especially in developing countries. Legumes belong to the family Leguminosae also known as Fabaceae whose seeds present in the pod represent a basic component of Indian diet since ancient times. Legumes like chickpea (*Cicer arietinum*), pigeon pea (Cajanus cajan), cowpea (*Vigna unguiculata*), mung bean (*Vigna radiata*), urdbean (*Vigna mungo*), lentil (*Lens culinaris*), French bean (*Phaseolus vulgaris*), horse gram (*Macrotyloma uniflorum*), field pea (*Pisum sativum*), moth bean (*Vigna aconitifolia*), lathyrus, soybeans (*Glycine max*), peanut (Arachis hypogaea) etc. serve an important role in the food chain, both for human and animal consumption as a source of rich protein (16–50 %) which are twice the amount found in whole grain cereals, dietary fibre, low in saturated fat, vitamins, no cholesterol and 15 essential minerals (Wang *et al,*. 2009) and it also contains 390-510 g kg^{-1} starch content as an important energy source. These are also potential future alternatives in production of bioenergy and alcoholics (Karlsson *et al.,* 2015; Lienhardt *et al.,* 2019). Pulses are an excellent source of a healthy diet, as they reduce the risk of cardiovascular diseases due to their low saturated fat and high Fiber content. It is reported that inclusion of pulses in our meals, reduced the risk to 80% of heart attacks, diabetes, strokes, and certain cancers, particularly colorectal cancer (Chibbar *et al.,* 2010). Pulse seeds can also enrich the blood quality, skin ailments, and ear inflammation.

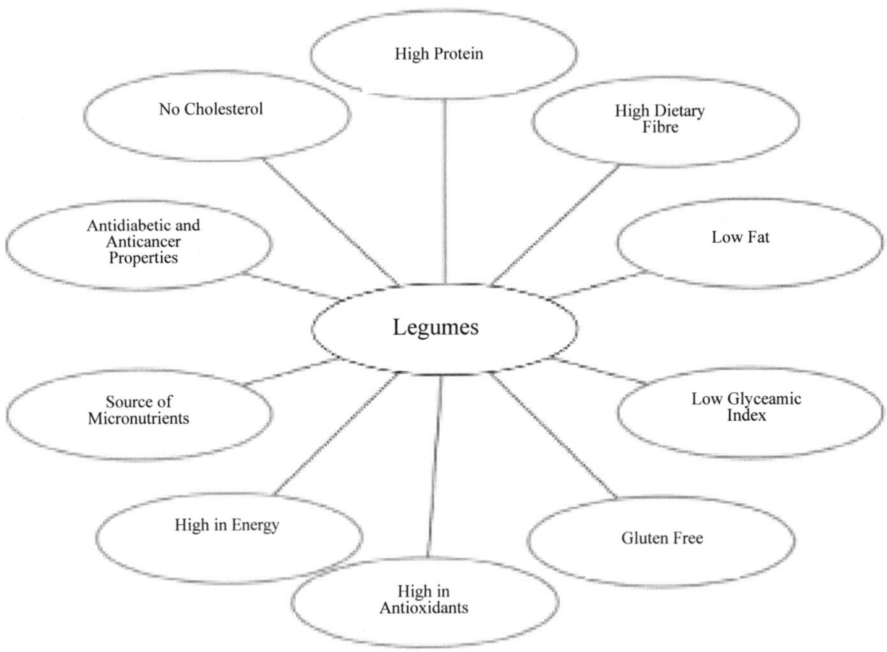

Fig. 14.1: Desirable attributes of legumes

Legumes are nutritious, affordable, and easily available food source being a part in the diet of millions of people globally, especially in areas where the cost of meat, dairy product, and fish are unaffordable (FAO, 2016). since the green revolution, the pulse-based diet has shifted to cereal-based diet. So, academic and governmental institutions are striving to emphasize the importance and benefits of legumes in the food system. To reach this goal, it's necessary to enhance their global production, incorporate legumes into farming systems, advance research on new varieties, create innovative management strategies, and overcome current market challenges (Bokern *et al.,* 2017). Food legumes are categorized into two groups: oil seeds and pulses. Oil seeds, like soybeans and peanuts, are rich in oil content, while pulses include all dry seeds of cultivated legumes used as traditional food. Legumes are ideal crops for farmers who cannot afford costly irrigation systems and fertilizers because it can thrive in poor soils and adverse weather conditions, exhibit higher resistance to diseases and pests, and serve as cover crops thereby reducing soil erosion, weeds, pests and diseases and improving soil structure (Watson *et al.,* 2017), green manure, catch crops in short season. Additionally, legumes form a symbiotic relationship with the nitrogen-(N)-fixing bacteria in their root nodules, converting atmospheric nitrogen into a usable form for plants which reduced the need for synthetic nitrogen fertilizers due to increasing soil fertility (Jensen *et al.,* 2012), thus making legumes excellent rotation crops (Kalidass and Mahapatra, 2014; Nedumaran *et al.,* 2015). In addition to saving farmers' production costs, it also reduces the environmental impact of agriculture by reducing nitrogen runoff, which may contaminate water and increase greenhouse gas emissions.

Now a days, a significant concern in worldwide legume production is the excessive accumulation of salts in the soil and the use of saline groundwater (Farooq *et al.,* 2017) and saline stress is known to disrupt the photosynthetic process and hormonal regulation, leading to nutritional imbalances in legumes, which in turn reduces yield and grain quality (Reckling *et al.,* 2016). One potential solution is to develop transgenic crops and implement management strategies to boost the yield of grain legumes in saline soils. Therefore, it is essential develop new varieties that are resistant to adverse conditions, thereby improving their nutritional value and optimizing their health benefits. Today, legumes can be seen as a superfood for the future, helping us to achieve zero hunger at a time when one in five children under the age of five is chronically malnourished (UNICEF, 2018). Despite their benefits, legume production remains lower than that of other staple crops like cereals, starchy roots, and vegetables. Therefore, it is essential to raise awareness about the benefits of legumes and promote their production. So, this chapter therefore attempts to create an extensive framework about the significance of grain legumes as a diversified diet in order to ensure food and nutritional security.

Statistics of Pulses

Globally, the major pulses during 2022 are grown over an area of 95.96 million ha with a production of 97.39 million metric tonnes with productivity 1015 kg ha^{-1} (Table 1.1). The major pulse-growing countries of the world include India (28 metric tonnes), Canada (6.1 metric tonnes), China (4.8 metric tonnes), Russian Federation (4.6 metric tonnes), Nigeria (4.2 metric tonnes), Australia (4.1 metric tonnes), Myanmar (3.6 metric tonnes) etc. (FAOSTAT, 2020). Pulses are grown by 170 countries around the world. Among different pulses, dry beans are the most widely cultivated by 105 countries pulses in the world, which contributed about 38% area to total world area and 29% share in production, Chickpea by 49 countries, contributed about 15% area and 19% share in production, Dry Peas by 95 countries, contributed 7% area and 15% share in production, Pigeon pea by 24 countries, contributed 6% area and 5% share in production, Lentil by 43 countries, contributed by 6% area and 7% share in production and others 11% in area and 15% share in production. In India, the major pulses during 2022 are grown over an area of 36.11 million ha with a production of 27.67 million metric tonnes with productivity 766 kg ha^{-1} (Table 1.2). It is clear that the country's productivity is 766 kg ha^{-1} is significantly lower than the global average of 1015 kg ha^{-1} but India is the largest area, producer as well as consumer of pulses with 38% global share in the area, 28 % share in the global production, consuming 27% of the world's pulses, and importing 14% of pulses in the world. Pulses cover approximately 23% of the total foodgrains area and contribute about 9-10% to the country's total foodgrains production. The top five pulses-producing states in India are Madhya Pradesh (22%), Maharashtra (16%), Rajasthan (16%), Uttar Pradesh (10%), and Karnataka (8%). Major *kharif* season pulse crops are arhar, urd, moong, lobia, kulthi and moth. Major *rabi* season pulse crops are gram, lentil, pea, lathyrus and rajma and major spring season pulse crops are urd, moong, lobia. Rabi pulses contribute over 60% to the total production. India has become self-reliant and a top producer of chickpeas (chana), pigeonpea and other pulse crops. Globally, the per capita intake of pulses is only about 21 grams per day. World Pulses Day is celebrated every year on February 10th, a tradition that began with the UN's International Year of Pulses in 2016.

Table 14.1: Area, Production and productivity of different pulses in the world

Crops	Area (m ha.)	Production (m. tonnes)	Yield (kg ha^{-1})
Chickpea	14.81	18.09	1222
Pigeonpea	6.03	5.33	883
Lentils	5.5	6.66	1209
Peas Dry	7.16	14.16	1979
Beans Dry	36.79	28.35	770
Cowpeas	15.19	9.78	643
Others	10.48	15.02	1434
Total Pulses	95.96	97.39	1015

Source: FAOSTAT, 2020

Table 14.2: Area, Production and productivity of different pulses in the world (2021-22)

Major Countries	Area (m ha.)	Production (m. tonnes)	Yield (kg ha^{-1})
India	36.11	27.67	766
Canada	3.28	6.16	1882
China	2.61	4.83	1848
Russian federation	2.31	4.61	1996
Nigeria	4.93	4.20	853
Australia	2.32	4.16	1787
Myanmar	3.80	3.58	941
Ethiopia	1.67	3.23	1938
Niger	6.03	2.96	491
Others	32.90	35.99	1094
Total Pulses	95.96	97.39	1015

Source: FAOSTAT, 2020

Malnutrition: an overview

Malnutrition is a serious condition that occurs when the body doesn't get the right amount of nutrients it needs to function. It can be caused by a deficiency, excess, or imbalance in a person's nutrient intake. Child malnutrition is a significant public health issue in numerous developing nations. According to the Food and Agriculture Organization (FAO), over 800 million people worldwide still suffer from hunger, and malnutrition affects nearly 2 billion people due to insufficient nutrient intake. Grain legumes, which are high in protein, micronutrients, and dietary fiber, can help combat these issues by improving both food and nutritional security. Beyond causing immense physical and emotional suffering, it severely hampers these countries' development prospects. Malnutrition, in all its forms, includes undernutrition (wasting, stunting, underweight), inadequate vitamins or minerals, overweight, obesity, and resulting diet-related noncommunicable diseases. As stated by WHO, 390 million adults were underweight and 2.5 billion overweight in 2022, including 890 million obese individuals. It was anticipated that 149 million children under the age of five were stunted (too little for their age), 45 million wasted (too thin for their height), and 37 million overweight or obese people worldwide in 2022. Undernutrition is a factor in about half of the deaths of children under the age of five. These primarily happen in nations with low and moderate incomes. The global burden of malnutrition has severe and long-lasting effects on people and their families, as well as on communities and entire nations in terms of development, economics, social issues, and medicine. The worldwide burden of malnutrition has detrimental effects on social, cultural, economic, and medical development. Worldwide in 2022 among children under five years of age, an estimated 148.1 million (22.3 percent) were stunted, 45 million (6.8 percent) were wasted and 37 million (5.6 percent) were overweight (Global Nutrition Report, 2023). Malnutrition refers to deficiencies, excesses, or imbalances in a person's intake of energy and/or nutrients. The term malnutrition addresses 3 broad groups of conditions:

- Undernutrition: which includes wasting (low weight-for-height), stunting (low height-for-age) and underweight (low weight-for-age).
- Overnutrition: overweight, obesity and diet-related noncommunicable diseases (such as heart disease, stroke, diabetes and some cancers).
- Micronutrient deficiency: malnutrition, which includes micronutrient deficiencies (a lack of important vitamins and minerals) or micronutrient excess.

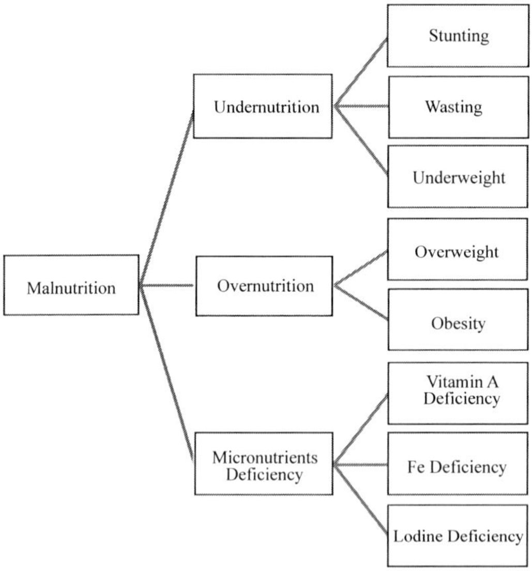

Fig. 14.2: Types of malnutrition (Adopted from WHO, 2024)

Food Security and Grain Legumes

Many lifestyle disorders arise from a poor diet that is high in animal products and low in plant materials. Legumes are rich in complex, low-glycemic carbohydrates, devoid of cholesterol, low in saturated fat, and abundant in bioactive compounds. These food elements improve health and lengthen life by reducing insulin production and preventing chronic illnesses such as diabetes, cancer, cardiovascular disease, and obesity. Consequently, eating a lot of beans can help you live a longer, healthier life.

- **Increased Yield Potential in Marginal Areas:** Grain legumes are cultivated in various agro-ecological zones, from semi-arid regions to high-altitude terrains. Due to their genetic diversity, legumes can be bred for higher yields and improved resilience in unfavourable conditions, such as drought-prone areas, where other crops often fail. For instance, cowpeas and pigeon peas are widely grown in sub-Saharan Africa for their drought tolerance, providing food security for millions in regions with irregular rainfall patterns. Additionally, legumes require minimal chemical inputs, which reduces farming costs. This feature is especially beneficial for smallholder farmers, who make up a significant portion of agricultural producers in developing countries. As per the International Crops Research Institute for the Semi-Arid Tropics (ICRISAT), legume varieties that withstand abiotic stress contribute to ensuring food security in marginal lands.

- **Enhancing Sustainable Agriculture via Nitrogen Fixation:** One of the most critical roles of legumes in sustainable agriculture is their ability to fix atmospheric nitrogen. Through symbiotic relationships with rhizobial bacteria in root nodules, legumes convert nitrogen from the atmosphere into a form usable by plants. This process not only enriches soil fertility but also decreases the need for synthetic nitrogen fertilizers. Studies indicate that legume crops can fix up to 200 kg of nitrogen per hectare per year, depending on the species.This nitrogen-fixing capacity improves the productivity of subsequent crops when legumes are used in crop rotation with cereals like wheat, rice, or maize. In countries like India, intercropping legumes with cereals has been shown to increase overall land productivity by up to 50% while also improving soil structure and water retention.
- **Legumes in Climate-Smart Agriculture:** Climate change poses a significant risk to food production systems globally, with increased temperatures, changing rainfall patterns, and more frequent extreme weather events. Grain legumes contribute to climate resilience due to their ability to thrive in dry, nutrient-poor soils. For example, pigeon pea and lentil are highly drought-tolerant, making them ideal crops for regions affected by water scarcity. Moreover, legumes have a low carbon footprint compared to many other crops due to their reduced reliance on nitrogen fertilizers. This characteristic aligns with sustainable development goals aimed at lowering agricultural greenhouse gas emissions.

Nutritional Security and Grain Legumes

- **High-Quality Protein Source:** Legumes are an outstanding source of high-quality protein, ranging from 20 - 45% protein which are particularly rich in the essential amino acid lysine (Philips, 1993). Peas and beans have a lower protein content (17 - 20%) while lupins and soybeans are at the higher end (38 - 45%) protein. As compared to most plant foods, legumes have almost double protein content than that of cereals (Kouris-Blazos and Belski, 2016). This high protein content is largely due to the nitrogen-fixing bacteria in their roots, which convert unusable nitrogen gas into ammonium, a form that the plant can use for protein synthesis. The primary components of leguminous protein are globulins and albumins, which are further subdivided into vialin and legumin. Vialin is the predominant protein group in most legumes, is notable for its low sulfur-containing amino acids (SCAA) content. Interestingly, this low SCAA content can be advantageous as it may enhance calcium retention. The breakdown of SCAA produces hydrogen ions, which can lead to bone demineralization and increased calcium excretion in urine. Consequently, leguminous proteins might improve calcium retention compared to high SCAA proteins from animal or cereal sources. Additionally, legume proteins help to reduce low-density lipoproteins (LDL), which are an important factor in the development of coronary heart disease (Philips, 1993). In terms of protein, cereals and legumes are complementary since cereals are high in SCAA which are poor in legumes and low in lysine which are high in legumes (Staniak *et al.,* 2014). Therefore, combining legumes with cereals significantly enhances protein quality. For a balanced diet, it is recommended to consume legumes and cereals in a 35:65 ratio. Legumes are vital in vegetarian diets as they are a primary source of protein along with essential vitamins and minerals. Protein-energy malnutrition is a widespread problem, particularly in regions where animal protein is inaccessible or unaffordable. Legumes are a vital plant-based source of protein, supplying essential amino acids, such as lysine, which is often deficient in cereal-based diets. According to the World Health Organization (WHO), legumes contain 18-25% protein, making them one of the best sources of plant-derived protein. Combining legumes with cereals (e.g., rice and beans) creates a complementary protein profile that provides all the essential amino acids needed for human nutrition. In India, for instance, pulses are a primary protein source for a significant portion of the population, helping mitigate protein deficiency-related issues like stunting and wasting in children.

- **Micronutrient-Rich Composition:** Legumes are also rich in micronutrients, which are vital for addressing hidden hunger-deficiencies in essential vitamins and minerals. Iron, zinc, and folate are among the most critical nutrients provided by grain legumes. These micronutrients help in preventing conditions like iron-deficiency anemia, Rana *et al.,* (2024), which affects nearly 30% of the global population, particularly women and children in developing regions. Legumes are a poor source of fat-soluble vitamins and vitamin C, but they are an excellent source of B-group vitamins like folate, thiamin, and riboflavin (Kouris-Blazos *et al.,* 2016). In addition to being a necessary vitamin, folate has been shown to lower the chance of neural tube abnormalities in newborns, such as spina bifida (Messina, 2016, FAO, 2016). Legumes are also sources of the essential minerals zinc, iron, calcium, selenium, phosphorus, copper, potassium, magnesium and chromium (Kouris-Blazos *et al.,* 2016). These micronutrients are essential for the maintenance of healthy bones (calcium), the metabolism of iron and carbohydrates (chromium, zinc), the production of haemoglobin (iron), antioxidative activity, protein synthesis, and plasma membrane(zinc) stabilization (Mogobe *et al.,* 2015). Legumes often have low salt content, which is beneficial considering current trends that support reducing sodium intake (Leonard, 2012) Despite having high iron levels, legumes are not a very effective source of iron due to their low bioavailability (Messina, 2016). Legumes can enhance iron absorption, though, when eaten in conjunction with meals high in vitamin C. Consequently, a high iron level would be crucial in preventing anaemia, particularly in fertile women.

The bioavailability of these nutrients can sometimes be limited by the presence of anti-nutritional factors, such as phytates, but various processing techniques like soaking, sprouting, and fermentation help reduce these compounds, making legumes an even more potent nutritional source.

- **Dietary Fiber and Health Benefits:** Legumes are also an excellent source of dietary fiber, which promotes digestive health and has been linked to reduced risks of cardiovascular disease, type 2 diabetes, and certain cancers. Studies have shown that legume-rich diets contribute to lower cholesterol levels and better blood sugar management, making them an ideal food for individuals with or at risk of diabetes. In fact, a study published in The American Journal of Clinical Nutrition found that consuming legumes at least three times per week was associated with a significant reduction in the risk of cardiovascular diseases by 22%. The total oligosaccharides in most legumes can reach 50 mg/g. Gas is a common side effect of eating beans and is caused by oligosaccharides. Since galactoside-containing oligosaccharides like raffinose and stachyose lack an α-galactosidase enzyme in the human gastrointestinal tract to cleave the α-1,6 galactose linkage, these oligosaccharides pass undigested to the colon where bacteria metabolize them to produce carbon dioxide, hydrogen, and methane in large amounts. Expelled from the body as flatulence, these gases can induce bloating and discomfort in the stomach. Notwithstanding the unfavorable perception around legume oligosaccharides, their positive characteristics exceed their drawbacks (Messina, 2016).
- **Legume consumption for livelihood:** Across the world, legumes constitute a staple in many diets, but in emerging and third-world nations in Asia, Latin America, and Africa, they are especially significant. As seen by the spread of consumption across various locations and the

found negative correlation between income and legume consumption, legumes have been named the "poor man's meat" (Messina, 2016). In some cases, this nickname appears to be accurate. But new studies are promoting legumes as "health food," pushing even the wealthiest individuals to include them in their diets Kouris-Blazos and Belski, 2016). Numerous commercial goods, including tofu, textured vegetable protein (TVP), soy sauce, soy paste, and curry, have been made from legumes. dietary fiber, single-cell proteins, citric acid, and enzymes are a few of the byproducts of legumes. As demonstrated in Table 2, there are numerous methods to include legumes in a diet that is well-balanced and healthful (Bouchenak and Lamri-Senhadji, 2013).

Table 14.2: Legumes are used as food by different means

Common name	Food uses
Peanut/Groundnut (*Arachis hypogaea*)	Peanut butter, peanut bar, flour, roasted/boiled snacks
Soybean (*Glycine max*)	Asian dishes (tofu, natto miso), roasted snacks, milk, yoghurt, sprouted beans, curd, yuba, soy sauce, soy paste, TVP
Black gram (*Vigna mungo*)	Dhal, fermented products (idli, dosa, papad)
Lentils (*Lens culinaris*)	Dhal, papadums
Peas (*Pisum sativum*)	Soup, dhal
Adzuki beans (*Vigna angularis*)	Japanese desserts and confections, soup ingredients for therapeutic purposes
Anasazi beans (*Phaseolus vulgaris*)	Boiled meal, snack, soup
Black-eyed peas (*Vigna unguiculata*)	Boiled snack/part of meal, fried cake *akara,* steamed pudding *moi moi* in West Africa
Chickpea (*Cicer arietinum*)	Middle Eastern and Mediterranean foods such as falafel and hummus, Boiled/fried/cooked/crushed snacks, dhal, curry, flour used in bread making, fermented food (*dhokla)*
Kidney beans (*Phaseolus vulgaris*)	Ingredient in Mexican chili; most-consumed legume in America
Lentils (*Lens culinaris*)	Daal, Soups and stews; most important legume in India
Lima beans (Phaseolus lunatus)	Cooked whole
Mung beans/Green gram (*Vigna radiate*)	Bean sprouts, cooked whole or with sugar into a dessert, soup, flour used for baking, transparent noodles, patties, sweets
Faba bean (*Vicia faba*)	Whole food

Source: Maphosa and Jideani, 2015; Murevanhema *et al.,* 2013; Qayyum *et al.,* 2012

Economic Empowerment and Poverty Reduction

- **Support for Smallholder Farmers:** Grain legumes offer significant economic opportunities, particularly for smallholder farmers. Since they are often grown on small plots of land and can be easily stored and traded, they provide a reliable income stream. According to FAO estimates, pulses contribute to the livelihoods of over 300 million smallholder farmers globally. Women, who form the backbone of many rural agricultural economies, are especially engaged in legume cultivation and processing. Moreover, growing legumes can reduce farmers' input costs by decreasing their reliance on chemical fertilizers, thanks to their nitrogen-fixing ability. This cost-saving makes legume cultivation an economically viable option for farmers in resource-poor regions.
- **Trade and Market Potential:** The global demand for pulses has been increasing steadily due to their nutritional benefits and the growing trend toward plant-based diets. Countries like India, which is the world's largest producer and consumer of pulses, have seen substantial trade in legumes, with imports supplementing domestic production to meet national demand. As the popularity of grain legumes grows in Western countries due to health-conscious and vegan/vegetarian diets, developing nations have the opportunity to capitalize on this global market.

Role in Alleviating Hidden Hunger

Grain legumes have been recognized as a strategic crop for combating hidden hunger, a condition where people consume enough calories but lack essential vitamins and minerals. In many countries, efforts are underway to biofortify legume crops to enhance their micronutrient content. For example, biofortified beans with higher iron content are being developed to address anemia in regions of Africa where iron deficiency is prevalent. Programs like HarvestPlus, in collaboration with CGIAR institutions, have spearheaded these initiatives to ensure nutrient-rich legume varieties reach vulnerable populations (Ranjan *et al.*, 2022). Underutilised legumes also known as orphan crops, neglected crops or lesser crops such as Bambara groundnut, African locust bean, African yam bean, pigeon pea, kidney bean, lima bean and marama bean deserve to be given more attention (Ebert, 2014). Most of these underutilised legumes thrive in adverse conditions, are nutritionally superior and yield more than common legumes (Ebert, 2014). The majority are grown mainly as secondary crops in households. To ensure that these legumes reach their full potential, great research should be done to increase our understanding of both the technical and practical aspects of these crops, Rana *et al.,* (2024). The substantial nutritious content of these legumes may help fight malnutrition. Underutilized legumes are thought to have a wealth of unidentified bioactive chemicals that could be used to create useful, economical, and medicinal foods. Increasing the usage of underutilized legumes could help decrease the overuse of common legumes like soybeans.

Conclusion

Grain legumes are critical for addressing both food and nutritional security on a global scale. Their high protein content, nutrient density, adaptability to diverse climates, and nitrogen-fixing ability make them indispensable in sustainable agriculture and food systems. Moreover, their role in promoting economic empowerment among smallholder farmers and reducing reliance on chemical inputs contributes to a more resilient and sustainable agricultural model. Governments, international organizations, and researchers must continue to promote the cultivation, biofortification, and consumption of grain legumes to achieve long-term food and nutritional security for all populations. Legumes' composition makes them appealing to health-conscious consumers, patients with celiac disease and diabetes, as well as consumers concerned with weight management. To harness the nutritional benefits of legumes, they should be incorporated into children's and infants' diets at home and through school feeding programs, especially in developing countries to reduce poverty and malnutrition. Legumes are a sustainable and affordable source of protein, unsaturated fat, dietary fiber, complex carbohydrates, micronutrients, and important bioactive phytochemicals; as such, their consumption could contribute to a healthier lifestyle.

References

Blössner, M., & de Onis, M. (2005). Malnutrition: Quantifying the health impact at national and local levels (WHO Environmental Burden of Disease Series, No. 12). World Health Organization.

Bouchenak, M., & Lamri-Senhadji, M. (2013). Nutritional quality of legumes, and their role in cardiometabolic risk prevention: A review. Journal of Medicinal Food, 16(3), 185-198. https://doi.org/10.1089/jmf.2011.0238

Chibbar, R. N., Ambigaipalan, P., & Hoover, R. (2010). Molecular diversity in pulse seed starch and complex carbohydrates and its role in human nutrition and health. Cereal Chemistry, 87, 342-352. https://doi.org/10.1094/CCHEM-87-4-0342

Ebert, A. W. (2014). Potential of underutilized traditional vegetables and legume crops to contribute to food and nutritional security, income, and more sustainable production systems. Sustainability, 6, 319-335. https://doi.org/10.3390/su6010319

FAO. (2016). Legumes can help fight climate change, hunger, and obesity in Latin America and the Caribbean. Food and Agriculture Organization of the United Nations and World Health Organization.

FAOSTAT. (2020). Food and Agriculture Organization of the United Nations. http://www.fao.org/faostat/en/

Farooq, M., Gogoi, N., Hussain, M., Barthakur, S. P., Bharadwaj, N., Migdadi, H. M., Alghamdi, S. S., & Siddique, K. H. M. (2017). Plant Physiology and Biochemistry, 118, 199-210.

Jensen, E. S., Peoples, M. B., Boddey, R. M., Gresshoff, P. M., Hauggaard-Nielsen, H., Alves, B. J. R., & Morrison, M. J. (2012). Legumes for mitigation of climate change and the provision of feedstock for biofuels and biorefineries: A review. Agronomy for Sustainable Development, 32, 329-364. https://doi.org/10.1007/s13593-011-0056-7

Kalidass, C., & Mahapatra, A. K. (2014). Evaluation of the proximate and phytochemical compositions of an underexploited legume Mucuna pruriens var. utilis (Wall ex Wight) L. H. Bailey. International Food Research Journal, 21(1), 303-308.

Karlsson, H., Ahlgren, S., Strid, I., & Hansson, P. A. (2015). Faba beans for biorefinery feedstock or feed? Greenhouse gas and energy balances of different applications. Agricultural Systems, 141, 138-148. https://doi.org/10.1016/j.agsy.2015.10.004

Khalid, I. I., & Elharadallou, S. B. (2013). Functional properties of cowpea (Vigna Ungiculata L. Walp) and lupin (Lupinus Termis) flour and protein isolates. Journal of Nutrition and Food Science, 3, 234. https://doi.org/10.4172/2155-9600.1000234

Kouris-Blazos, A., & Belski, R. (2016). Health benefits of legumes and pulses with a focus on Australian sweet lupins. Asian Pacific Journal of Clinical Nutrition, 21(1), 1-17. https://doi.org/10.6133/apjcn.2016.25.1.23

Leonard, E. (2012). Cultivating good health. In Grains and Legumes Nutrition Council (pp. 3-18). Cadillac Printing.

Lienhardt, T., Black, K., Saget, S., Porto Costa, M., Chadwick, D., Rees, R. M., Williams, M., Spillane, C., Iannetta, P. M., Walker, G., & Styles, D. (2019). Just the tonic! Legume biorefining for alcohol has the potential to reduce Europe's protein deficit and mitigate climate change. Environment International, 130, Article 104870. https://doi.org/10.1016/j.envint.2019.104870

Maphosa, Y., & Jideani, V. A. (2015). Dietary fiber extraction for human nutrition: A review. Food Reviews International, 32(1), 98-115. https://doi.org/10.1080/87559129.2015.1057840

Messina, M. J. (2016). Legumes and soybeans: Overview of their nutritional profiles and health effects. Asia Pacific Journal of Clinical Nutrition, 25(1), 1-17. https://doi.org/10.1.1.847.8636

Mogobe, O., Mosepele, K., & Masa, W. R. L. (2015). Essential mineral content of common fish species in Chanoga, Okavango Delta, Botswana. African Journal of Food Science, 9(9), 480-486. https://doi.org/10.5897/AJFS2015.1307

Murevanhema, Y. Y., & Jideani, V. A. (2013). Potential of bambara groundnut (Vigna subterranea (L.) Verdc) milk as a probiotic beverage: A review. Critical Reviews in Food Science and Nutrition, 53(9), 954-967. https://doi.org/10.1080/10408398.2011.574803

Nedumaran, S., Abinaya, P., Jyosthnaa, P., Shraavya, B., Parthasarathy, R., & Bantilan, C. (2015). Grain legumes production, consumption and trade trends in developing countries (Working Paper Series No. 60). International Crops Research Institute for the Semi-Arid Tropics (ICRISAT). https://doi.org/10.7910/DVN/V61SNB

Qayyum, M. M. N., Butt, M. S., Anjum, F. M., & Nawaz, H. (2012). Composition analysis of some selected legumes for protein isolates recovery. The Journal of Animal & Plant Sciences, 22(4), 1156-1162.

Rana, L., Kumar, N., Rajput, J., Sow, S., Ranjan, S., Kumari, S., Pradhan, J., Kumar, A., Singh, S.N., Kumar, A., Jha, C.K., Kumari, M., Singh, D., Sahoo, R., 2024. Unlocking potential: The role of zinc fortification combating hidden hunger and enhancing nutritional security. J. Exp. Agric. Int. 46(10), 625–642. https://doi.org/10.9734/jeai/2024/v46i102986.

Ranjan, S., Sow, S., Kumar, S., Ghosh, M., Roy, D. K., Dutta, S. K., & Pramanick, B. (2022). Enhancing Nutrient Use Efficiency and Productivity of Cereals through Site Specific Nutrient Management. Journal of Cereal Research, 14(3), 229-242.

Reckling, M., Bergkvist, G., Watson, C. A., Stoddard, F. L., Zander, P. M., Walker, R. L., Pristeri, A., Toncea, I., & Bachinger, J. (2016). Frontiers in Plant Science, 7, 669. https://doi.org/10.3389/fpls.2016.00669

Sow, S., & Ranjan, S. (2021). A Review on agronomic biofortification for improving food and nutritional security. Research Biotica, 3(2), 139-44. https://doi.org/10.54083/ResBio/3.2.2021.139-144.

UNICEF/WHO. (2018). Levels and trends in child malnutrition. United Nations Children's Fund, the World Health Organization, and World Bank Group.

United Nations. (2013). World population prospects: The 2012 revision, key findings and advance tables (Working paper no. ESA/P/WP.227). Department of Economic and Social Affairs, Population Division, New York.

Wang, N., Hatcher, D. W., Toews, R., & Gawalko, E. J. (2009). Influence of cooking and dehulling on nutritional composition of several varieties of lentils (Lens culinaris). Food Science and Technology, 42, 842-848. https://doi.org/10.1016/j.lwt.2008.11.002

Watson, C. A., Reckling, M., Preissel, S., Bachinger, J., Bergkvist, G., Kuhlman, T., Lindström, K., Nemecek, T., Topp, C., Vanhatalo, A., Zander, P., Murphy-Bokern, D., & Stoddard, F. (2017). Grain legume production and use in European agricultural systems. Advances in Agronomy, 144, 235-303. https://doi.org/10.1016/bs.agron.2017.03.003

Webb, P., Stordalen, G. A., Singh, S., Ramani, W. B., Prakash, S., Anna, L., et al. (2018). Hunger and malnutrition in the 21st century. British Medical Journal, 361, k2238. https://doi.org/10.1136/bmj.k2238

15

Supplementing Diet Through Grain Legumes to Achieve Food Security

Bhagyashri Nandkumar Solanke*

Department of Agronomy, Dr. Panjabrao Deshmukh Krishi Vidyapeeth, Akola Maharashtra, India

**Corresponding author's email: bhagyasolanke99@gmail.com*

Abstract

Food insecurity and malnutrition are on the rise globally, impacting millions, especially in rural regions where extreme poverty and food insecurity are deeply rooted. Vulnerable groups, including women, youth, and indigenous peoples, are disproportionately affected. Producing food that meets dietary needs and ensuring its availability in the future is essential for sustainable development and food sustainability. Addressing the growing global food security concerns and supporting the increasing population requires the continual expansion of plant-based proteins' significance. Legumes are excellent sources of proteins, carbohydrates, and both soluble and insoluble dietary fibers. Legumes, with their high dietary fiber and low-fat content, help maintain body weight and lower the risk of cardiovascular disease. They are also rich in essential minerals such as potassium, calcium, magnesium, phosphorus, and iron, and provide important vitamins like folate, thiamine (B_1), riboflavin (B_2), and niacin (B_3). Additionally, pulses contain various phytochemicals, including phenolics, flavonoids, phytates, lectins, tannins, saponins, oxalates, enzyme inhibitors, phytosterols, and antimicrobial peptides, which offer numerous potential health benefits. These phytochemicals exhibit anti-inflammatory, anticancer, antimicrobial, and anti-ulcerative properties. Incorporating legumes into agricultural systems can foster more sustainable and climate-resilient farming practices. In developing countries, legumes are often used to supplement cereals or starchy foods to combat protein-energy malnutrition, as animal proteins tend to be expensive in these regions.

Keywords: *Antioxidants, Food security, Legume, Malnutrition, Healthy diet*

Introduction

The concept of food security refers to the state in which every person has consistent access to enough safe and healthy food that meets their dietary needs and preferences for an active and healthy life. This includes ensuring food availability, as well as economic and physical access to food, proper food utilization, and stability over time. Nutrition security ensures that everyone consistently has fair access to nutritious, safe, and affordable foods, which are vital for achieving and maintaining optimal health and well-being.

Proper nutrition is both a contributor to and an indicator of progress. The positive effects of adequate nutrition extend widely, impacting not only families and communities, but also regions and countries. Conversely, malnutrition obstructs national advancement and severely undermines the health, progress, and happiness of current and future generations. Malnutrition encompasses various forms of deficiencies, including undernutrition, micronutrient deficiencies, as well as overweight and obesity. The eradication of malnutrition is essential for the realization of the Sustainable Development Goals, especially SDG 2 (Zero Hunger), SDG 3 (Good Health and Well-being), and SDG 10 (Reduced Inequalities).

The responsibility of achieving the Sustainable Development Goals (SDGs) falls upon all nations. Despite progress in populous countries with growing economies, many nations worldwide are still facing increasing hunger, food insecurity, and malnutrition. This trend affects millions, especially in rural areas where extreme poverty and food insecurity are prevalent. Vulnerable groups, such as women, youth, and Indigenous Peoples, are disproportionately impacted by these challenges. The factors driving food insecurity and malnutrition worldwide include conflict, climate variability and extremes, economic slowdowns and downturns, lack of access to and affordability of healthy diets, unhealthy food environments, and high and persistent inequality.

The 2023 assessment of global hunger, based on the prevalence of undernourishment (PoU) (SDG Indicator 2.1.1), shows a persistent lack of progress towards achieving Zero Hunger. After a significant increase from 2019 to 2021, the proportion of the global population experiencing hunger has remained nearly unchanged for three years, with the latest estimates indicating a PoU of 9.1 percent in 2023. This translates to between 713 and 757 million people (8.9 to 9.4 percent of the global population) being undernourished in 2023. Using the mid-range estimate of 733 million, approximately 152 million more people faced hunger in 2023 compared to 2019. (FAO, 2024)

Healthy Diet

Food prices continued to rise in 2022 compared to the period from 2017 to 2021, leading to an increase in the average cost of a healthy diet globally and across all regions. The FAO Food Price Index surged by 52% between 2019 and 2022, with cereal prices rising by 60%, dairy products by 45%, meat by 19%, and oils by an astonishing 125% compared to pre-COVID-19 levels. In Asia, 1.66 billion people were unable to afford a healthy diet in 2022. However, this marked two consecutive years of improvement, with 163 million fewer people unable to afford a healthy diet in 2022 compared to 2020. (FAO, 2024).

Malnutrition

Malnutrition, including deficiencies in essential micronutrients, during critical stages such as foetal development, infancy, and early childhood, significantly increases the risk of illness and death. It hampers physical growth and weakens the immune system, leading to frequent infections. Additionally, it can cause suboptimal cognitive development and permanent changes in organ systems, predisposing individuals to chronic diseases in adulthood. Research indicates that prenatal nutrition interventions can enhance birth outcomes, which are linked to better educational achievements and human capital development later in life.

Children who consume nutritious foods during early childhood tend to be more economically productive as adults. Conversely, nutritional deficiencies during this period can impair brain development, affect learning abilities, and reduce school readiness, thereby limiting lifelong achievement potential and worsening health disparities and social inequality. The double burden of malnutrition, characterized by the simultaneous presence of undernutrition and overweight/obesity, has increased across all age and income groups in recent decades.

What are Legumes?

Legumes, belonging to the family Fabaceae or Leguminosae, are the edible seeds of leguminous plants. This family includes approximately 670-750 genera and 18,000-19,000 species. "Food legumes" encompass a variety of common legumes such as chickpeas (*Cicer arietinum*), mung beans (*Vigna radiata*), soybeans (*Glycine max*), peanuts (*Arachis hypogaea*), lentils (*Lens culinaris*), cowpeas (*Vigna unguiculata*), moth beans (*Vigna aconitifolia*), peas (*Pisum sativum*), and faba beans (*Vicia faba*).

Food legumes are categorized into two groups: pulses and oilseeds. Pulses are dry seeds consumed as food, while oilseed legumes are those with a notably high oil content. Legumes are resilient to drought and can thrive in marginal soils and climates. They have the unique ability to absorb nitrogen from the air and fix it into the soil (Nieuwenhuis and Nieuwelink, 2005). This nitrogen-fixing capability makes legumes crucial in tropical regions where soils often lack sufficient nitrogen. Cultivating legumes is a cost-effective and straightforward method for farmers to enrich the soil and reduce expenses on artificial fertilizers. Legumes have the potential to alleviate poverty among smallholder farmers, especially during droughts and increased weather variability, which have been prevalent in Africa and other developing countries (Mubaiwa *et al.*, 2017). Additionally, legumes play a vital role in providing nutrition and income for low-income rural populations in developing countries.

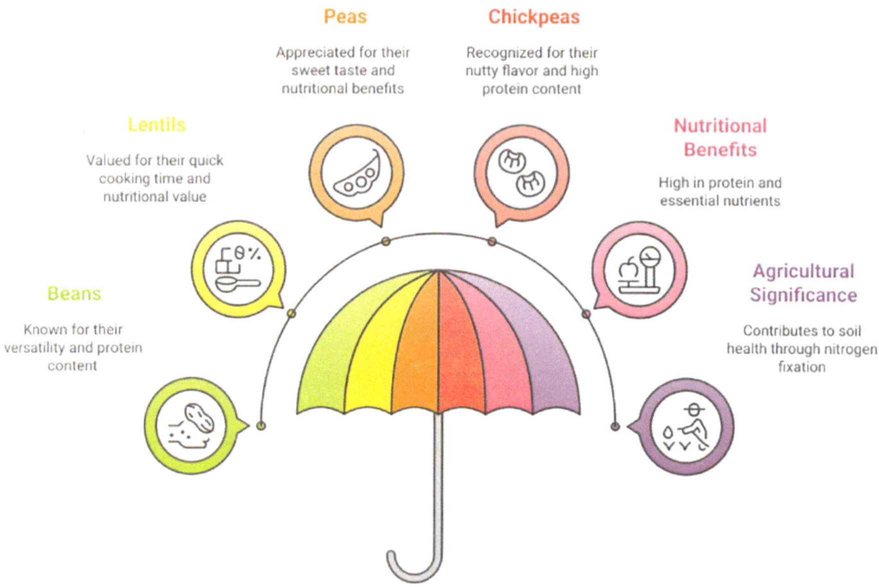

Fig. 15.1: Types of legumes

Around the world, legumes are used in a variety of ways, including flour, dehulled split grains, and whole seeds. The regions with the highest daily intake of legumes are South Asia, Sub-Saharan Africa, Latin America, and the Caribbean (33, 33, and 34g, respectively), whereas the regions with the lowest consumption are the Caucasus and Central Asia (1g/day) (Rawal and Navarro, 2019). A significant part of the global population depends on legumes as a dietary staple. This is particularly true in developing countries where meat, dairy products, and fish are prohibitively expensive. Due to their high protein content and affordability, pulses are often called "poor people's meat." However, recent research advocates for their inclusion in the diets of even wealthy individuals, leading to their recognition as "health food." Legumes are essential in the human diet due to their rich content of bioactive molecules, providing numerous nutritional and physiological benefits (Ranjan *et al.*, 2023).

They are packed with proteins, carbohydrates, dietary fiber, vitamins, and valuable phytochemicals that exhibit significant biological activities. Protein-rich pulses are a staple food for a large portion of the Indian population, meeting their protein and energy needs. Thanks to their high dietary fiber and low-fat content, legumes help maintain body weight and reduce the risk of cardiovascular diseases. They are also excellent sources of minerals such as potassium, calcium, magnesium, phosphorus, and iron, as well as vitamins like folate, thiamine (vitamin B_1), riboflavin (vitamin B_2), and niacin (vitamin B_3). Moreover, legumes contain various phytochemicals, including phenolics, flavonoids, phytates, lectins, tannins, saponins, oxalates, enzyme inhibitors, phytosterols, and antimicrobial peptides. These compounds offer potential health benefits, such as anti-inflammatory, anticancer, antimicrobial, and anti-ulcerative effects.

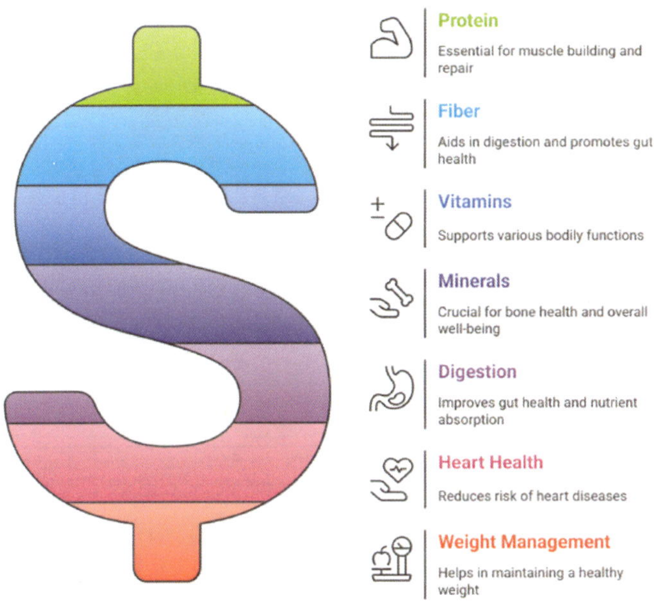

Fig. 15.2: Legumes and its benefits

Legume Area, Yield and Production: Global Context

According to the 2013 report from the United Nations, Department of Economic and Social Affairs, Population Division in New York, the global population is projected to reach 9.6 billion by 2050. Consequently, there will be an increased demand for legumes, leading to the necessity for higher food legume production. Since 1980, the global production of various grain legume crops, including chickpea, pigeonpea, cowpea, dry bean, faba bean, lentil, soybean, and groundnut, has been increasing at a rate of over 1% per year. Cowpea production has seen the highest growth, with an annual increase of 4.7% between 1994-95 and 2008-10, as new regions have started cultivating it, despite low yield levels.

Soybean cultivation has expanded rapidly in Latin America and South Asia, resulting in a production growth rate of 4.3% per year. Additionally, lentil and pigeonpea production have grown at annual rates of 2.2% and 2.0%, respectively, over the past 15 years. Soybean holds the largest market share among various legume crops, accounting for over 70%, followed by groundnut at 12%, dry bean at 7%, chickpea at 4%, and pea at 3.5%.

Table 15.1: Global grain legumes area, production and yield (Nedumaran *et al.*, 2015)

Particulars	**Drybean**	**Chickpea**	**Cowpea**	**Pigeon-pea**	**Lentil**	**Faba bean**	**Soybean**	**Groun-dnut**
Area harvested (million ha)								
1994–96	25.9	10.9	8.2	4.1	3.4	2.3	62.0	22.1
2008–10	27.4	11.5	10.4	4.7	3.7	2.5	99.4	24.0
% change	6	5	27	15	10	8	60	9
Growth rate (%)	0.7	0.2	1.7	1.1	0.8	0.6	3.5	0.4
Production (million tons)								
1994–96	17.1	8.1	2.7	2.8	2.8	3.6	131.2	29.4
2008–10	21.7	9.9	5.3	3.7	3.7	4.2	238.6	37.4
% change	27	23	91	35	35	18	82	27
Growth rate	1.9	1.3	4.7	2.1	2.3	1.4	4.4	1.7
Yield (kg/ha)								
1994–96	659	739	339	684	822	1,565	2,115	1,331
2008–10	793	864	508	800	1,003	1,701	2,399	1,558
% change	20	17	50	17	22	9	13	17
Growth rate	1.2	1.1	2.9	1.0	1.5	0.8	0.9	1.3

(Note: Percentage change and growth rate pertains to the period between 1994 and 2010. Also, growth rate is an annual figure)

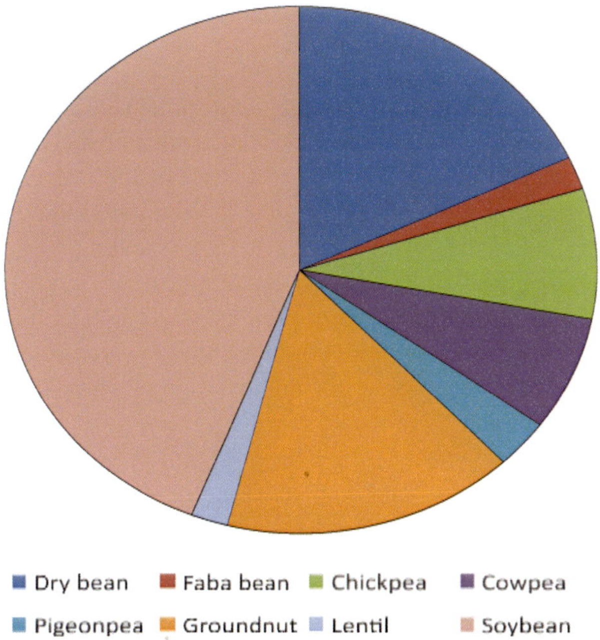

Fig. 15.1: Area shares (in %) of legumes developing world in 2008–10

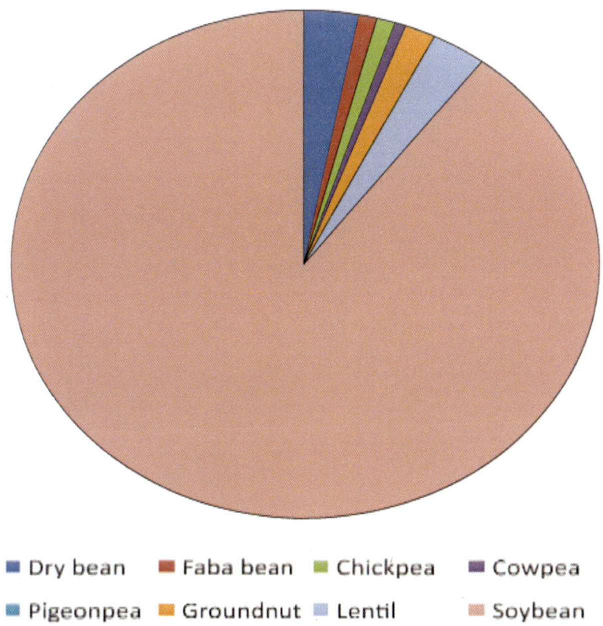

Fig. 15.2: Area shares (in %) of legumes in the in the developed world in 2008–10 (*Source:* Nedumaran *et al.*, 2015)

Nutrient components of legumes

a) Protein

The National Academy of Medicine advises that adults consume at least 0.8 grams of protein per kilogram of body weight daily. Legumes are an excellent source of high-quality protein, containing 20-45% protein with high levels of lysine but lower levels of sulfur-containing amino acids like methionine, cystine, and cysteine. Lupins and soybeans are especially rich in protein, with concentrations ranging from 35-45%. Legume seed proteins can be divided into storage, structural, and physiologically active proteins, such as lectins, enzymes, and enzyme inhibitors. Based on their solubility, these proteins are classified into albumins, globulins, prolamins, and glutelins, each of which can be extracted using different solutions.

b) Carbohydrates

Legumes are a rich source of complex carbohydrates, comprising up to 60% of their dry weight. The primary components are starch (35-55%), along with monosaccharides, oligosaccharides, and non-starch polysaccharides. Unlike starches from cereals and tubers, leguminous starch is digested more slowly. This results in a low glycemic index (GI), making legumes beneficial for blood glucose control, particularly for diabetic patients and those at risk of developing diabetes. Additionally, legumes are gluten-free, making them suitable for individuals with celiac disease or gluten sensitivity.

Table 15.2: Protein, carbohydrates and other nutrient components of different legumes (g per 100g of sample)

Legumes	Energy (Cal)	Protein	Carbohydrate	Fiber	Sugar	Fat	SFA	MUFA	PUFA	Ash	Water
Adzukibean	329	19.87	62.90	12.70	-	0.53	0.19	0.05	0.11	3.26	13.44
Chickpea	378	20.47	62.95	12.20	10.70	6.04	0.60	1.37	2.73	2.85	7.68
Cowpeas	336	23.52	60.03	10.60	6.90	1.26	0.33	0.10	0.54	3.24	11.95
Fababean	341	26.12	58.29	25.00	5.70	1.53	0.25	0.30	0.62	3.08	10.90
Frenchbean	343	18.81	64.11	25.20	-	2.02	0.22	0.13	1.20	4.30	10.77
Greengram	347	22.53	62.62	16.30	6.60	1.15	0.34	0.16	0.38	3.32	9.05
Kidneybean	333	22.53	60.01	24.90	2.23	0.83	0.12	0.06	0.45	3.83	11.75
Lablab bean	344	23.9	60.74	25.60	-	1.69	0.28	0.07	0.71	4.29	9.38
Lentil	352	24.63	63.35	10.70	2.03	1.06	0.15	0.19	0.52	2.71	8.26
Limabean	338	21.46	63.38	19.00	8.50	0.69	0.16	0.06	0.30	4.30	10.17
Lupin	371	36.17	40.37	18.90	-	9.74	1.15	3.94	2.43	3.28	10.44
Mothbean	343	22.94	61.52	-	-	1.61	0.36	0.12	0.75	4.26	9.68
Pea	364	23.12	61.63	22.20	3.14	3.89	0.40	0.61	1.02	2.67	8.69
Peanut	567	25.8	16.13	8.50	4.72	49.24	6.27	24.42	15.55	2.33	6.50
Pigeonpea	434	21.7	62.78	15.00	-	1.49	0.33	0.01	0.81	3.45	10.59
Soybean	446	36.49	30.00	9.30	7.33	20.00	2.88	4.40	11.25	4.87	8.54

(SFA- Saturated Fatty Acids, Mono- and polyunsaturated fatty acids (MUFA and PUFA); *Source:* USDA (United States Department of Agriculture) Standard Reference

Overall, legumes are valuable for those aiming for a healthy, disease-free lifestyle. In the food industry, legume starch isolates are commonly used as thickeners in soups and gravies. Legumes are also rich in resistant starch and oligosaccharides, such as raffinose, which have prebiotic properties. These compounds are fermented by probiotics into short-chain fatty acids, enhancing colonic health and reducing the risk of colon cancer.

c) Fibers

Children and adults need at least 25 to 35 grams of fiber daily for optimal health. Legumes are also high in dietary fiber, including cellulose, hemicellulose, lignin, and β-glucans. Dietary fiber is crucial for food digestibility and nutrient bioavailability. Soluble dietary fiber in legumes ranges from 0.0-4.6%, while insoluble fiber ranges from 13-30%. Diets high in dietary fiber offer numerous health benefits, including the prevention and potential treatment of conditions like constipation, obesity, diabetes, heart complications, hemorrhoids, and certain cancers. Soluble dietary fiber, in particular, can lower blood cholesterol, improve glucose tolerance, and reduce glycemic response by forming a protective gel along the intestinal walls, thereby reducing the absorption of glucose and cholesterol into the bloodstream.

Insoluble dietary fibers, being porous and low in density, increase fecal bulk and promote regular bowel movements. Consequently, legumes are an essential part of a healthy diet. In the food industry, dietary fiber fractions from legumes are utilized as stabilizers, texturing agents, fortifiers, bulking agents, fat replacers, and emulsion stabilizers in bakery, meat, extruded products, and beverages. Additionally, legumes are good sources of resistant starch, which functions similarly to dietary fiber in human nutrition. Increased consumption of fiber-rich legumes can offer various health benefits, including preventing weight gain in adults with high BMIs, mitigating the negative effects of being overweight, and reducing blood pressure (WHO, 2020).

d) Lipid

Legumes contain significant amounts of mono- and polyunsaturated fatty acids (PUFAs) and very few saturated fatty acids. The primary fatty acids in legumes are linoleic, oleic, palmitic, and linolenic acids. Legume seeds generally have low fat content. The primary fatty acids in legumes are linoleic, oleic, palmitic, and linolenic acids. Among legumes, chickpeas have the highest linoleic acid (a polyunsaturated fat) content, followed by soybeans and lentils. However, soybeans contain the highest levels of oleic acid (a monounsaturated fat) compared to chickpeas, lentils, and other legumes.

Cowpeas and mung beans are richer in omega-3 fatty acid (linolenic acid) than soybeans, chickpeas, and other legumes, while soybeans have the highest omega-6 fatty acid (linoleic acid) content. Although omega-3 fatty acids are known to inhibit platelet aggregation and reduce inflammatory cytokine production, potentially lowering the risk of cardiovascular disease, the levels in legumes may be too low to have a significant effect. Nonetheless, legumes can complement other foods that are higher in omega-3 fatty acids. The American Heart Association recommends that 8-10% of daily calories should come from polyunsaturated fats, and research indicates that increasing polyunsaturated fat intake to 15% of daily calories, replacing saturated fats, can reduce the risk of heart disease. Legume seeds, except for peanuts (45%), chickpeas (15%), and soybeans (47%), are generally low in fat, with only about 5% of their energy derived from fat.

e) Minerals

Legumes contain essential minerals such as calcium (32-394 mg/100 g), phosphorus (203-800 mg/100 g), magnesium (58-472 mg/100 g), iron (3.2-10 mg/100 g), zinc (1.6-6.3 mg/100 g), sodium (3-41 mg/100 g), and potassium (616-2300 mg/100 g). The mineral content in legumes varies based on the type and cultivar. Winged beans have the highest calcium content (80-370 mg/100 g), followed by chickpeas (57-197 mg/100 g) and soybeans (180 mg/100 g). Lentils (73-120 mg/100 g) and lupins (84 mg/100 g) have the lowest calcium levels. All legumes are rich in phosphorus. Soybeans have the highest potassium (1800 mg/100 g) and magnesium (280 mg/100 g) content.

Winged beans contain the most iron (2-18 mg/100 g), while chickpeas have the highest zinc content (1-6.8 mg/100 g). Since cereals and tuber-based diets are typically low in iron and zinc, adding legumes to these foods can enhance their mineral content. These micronutrients play crucial physiological roles, including supporting bone health (calcium), enzyme activity and iron metabolism (copper), carbohydrate and lipid metabolism (chromium, zinc), hemoglobin synthesis (iron), antioxidative activity, protein synthesis, and plasma membrane stabilization (zinc).

f) Vitamins

Vitamins are essential micromolecules needed in small but sufficient amounts to maintain optimal human health. Legumes are rich in various vitamins, including thiamine (0.3-1.6 mg per 100 g), riboflavin (0.12-0.33 mg per 100 g), niacin (traces-4.7 mg per 100 g), and folic acid (0.4-2.1 mg per 100 g). Lupins are particularly high in vitamin B_1 (thiamin) and B_2 (riboflavin), soybeans have the highest levels of vitamin B_3 (niacin) and folate, lentils are rich in vitamin

B_6 (pyridoxine), and chickpeas are notable for their vitamin C (ascorbic acid) content. Thiamine levels in most legumes meet the Recommended Dietary Allowance (RDA) for both children and adults. Mung beans, winged beans, and soybeans have riboflavin levels within the RDA for children, while lupins meet the RDA for adolescents. Chickpeas and lentils provide pyridoxine within the RDA range for children.

Although legumes are not major sources of vitamin C, chickpeas contain enough ascorbic acid to meet children's RDA, and both chickpeas and lentils have folate levels within the RDA for children. Soybeans provide folate within the RDA for adults. This information indicates that most legumes can fulfil the B-vitamin requirements for children and complement the vitamin intake for adults.

g) Antioxidants

In addition to their high protein and mineral content, legumes are rich in phenolic compounds, which are natural antioxidants and important bioactive compounds in foods. These phenolic compounds, which include plant secondary metabolites like gallic acid, have significant antioxidant capacities (4120 mg/L). Among the legumes studied, lupins have the lowest total phenolic content (0.19 mg/g), while lentils have the lowest flavonoid content (2.21 mg/g). The variation in phenolic compounds among legumes can affect their ability to scavenge free radicals. Tocopherols also contribute to the antioxidant activity of legumes.

Legumes contain four major tocopherol isomers (α-, β-, γ-, and δ-tocopherol). Although the levels of tocopherols in legumes are generally low, they are higher than in cereals. γ-tocopherol, in particular, has been noted for its protective effects against cardiovascular disease and its role in enhancing the shelf-life stability of legume seed oil due to its antioxidant properties. The average γ-tocopherol content in legumes ranges from 0.43 mg/100 g in cowpeas to 23.8 mg/100 g in soybeans, indicating that tocopherol content varies by legume species and variety.

h) Other Health Benefits of Legumes

Additionally, incorporating legumes into the diet may help prevent cardiovascular diseases and obesity while exhibiting anticancer properties. The cholesterol-lowering effect of legumes is attributed to the reduction of fatty acid production by fermentation by-products like propionate, acetate, and butyrate (Crujeiras *et al.*, 2007). When consumed, legumes form a viscous gel in the small intestine, slowing digestion and delaying hunger onset (McCrory *et al.*, 2010). Legumes have also been found to lower the risk of developing cancer, potentially by inhibiting aging and cancer-related pathways. Studies indicate that individuals who consume more legumes and lentils have a decreased risk of dying from cancer (Papandreou *et al.*, 2019). Specific components in legumes, such as lycopene, contribute to these health benefits.

Chickpeas and butyrate have been linked to reduced incidences of prostate and colorectal cancer, respectively. Chickpeas contain linoleic acid, which is used to synthesize prostaglandins. These prostaglandins can lower blood pressure and relax smooth muscles, potentially reducing the risk of coronary heart disease. Phytosterols in chickpeas also inhibit cholesterol absorption in the intestine, leading to lower levels of low-density lipoprotein cholesterol (LDL-C) in the blood (Pittaway *et al.*, 2007). Additionally, β-sitosterol in chickpeas helps reduce serum cholesterol and the risk of coronary heart disease. Legumes are particularly relevant as a tasty and healthy alternative to high-calorie meals common in developed nations. Incorporating legumes into a balanced diet can enhance overall well-being and address prevalent health concerns in developed societies.

Role of Grain Legumes for Achieving Food and Nutritional Security

Diets based on legumes have the potential to provide food and nutritional security for the world's rapidly growing population. According to Joshi and Rao (2017), wealthy nations primarily obtain their dietary protein from animal sources, whereas underdeveloped nations primarily depend on legume grains. In a large portion of the world, legumes are acknowledged as the most significant source of plant-based protein. A serving of cooked beans (140–190 g) provides less than 300 calories and a significant amount of daily recommended nutrients, such as up to 50% folate, 85% nutritious fibers, 15% protein, and potassium, according to the Swedish Food Composition Database (NFA, 2017). They are therefore the most solid option among diets that don't rely much on meat due to their exceptional nutritional profile.

Legumes are a solid source of protein to assist fulfil daily requirements, but they also contribute to muscle building and cell replacement and repair as a proteinaceous source. Known as the building blocks of the body, proteins are essential to daily diets in order to preserve the nutritional balance of the human organ system and other animals that humans have raised. Numerous micronutrients, including magnesium, phosphorus, potassium, zinc, and copper, are found in legumes and have a number of health benefits. Legumes are the only source of protein in economies that are expanding since a portion of the poor cannot afford to eat a diet high in animal protein. After being dehulled, beans are eaten as a dietary supplement in south Asia. As a result, several micronutrients found in the bean shells were lost. Additionally, this process depletes their nutrition of some micronutrients, which may cause problems for them. Cereals and pulses should therefore be included in the human diet in equal amounts to help with such issues.

As a result, combining the benefits of both the legume-oriented diet and the cereal-based diet-cereals being rich in lysine and tryptophan, while pulses are rich in methionine, cystine, and tryptophan-makes for a complete diet that includes all the essential nutrients. Remarkably, starch digestibility has also been shown to marginally improve in diets containing both wheat and a starch supply from legumes. Because they are less expensive than meat, legumes are also known as "poor man's meat" (Sharif *et al.*, 2018). Cereals are considered a staple meal globally and are an essential component of the human diet. But their poor protein content limits their use to just that. Complementary foods, such as grains and legumes, are being taken to make up for this protein deficit (Suri *et al.*, 2014). As an illustration, legumes deficient in methionine and cysteine can easily be added to cereals poor in lysine (Jiddere and Filli, 2015).

Additionally, regarded as an affordable and healthful source of fats are legumes. When it comes to cooking oils and fat, the majority of the market is either taken up by cheap palm oil, whose cultivation is deemed unsustainable and whose consumption is occasionally viewed as unhealthy (Aguiar *et al.*, 2018), or by canola oil, which has a positive impact on health but is expensive and difficult for those with low incomes to obtain. On the other hand, animal fat is used as cooking oil in many regions of the world, but its limited availability and health risks limit its use to a small class of people. To address these concerns, researchers have focused on legume crops with elevated fat content, such as soybeans and groundnuts. These legumes offer a healthy, sustainable, and cost-effective alternative to traditional cooking oils (Sow and Ranjan, 2021).

Moreover, the depletion of nutrients in mainstream food crops due to exhausted agricultural lands has led to a phenomenon known as "hidden hunger." This global micronutrient malnutrition affects many developing countries, resulting in health impacts. Food legumes, rich in essential micronutrients like zinc and iron, can play a vital role in combating hidden hunger and preventing disorders like anaemia. By promoting the consumption of legumes and implementing sustainable agricultural practices, we can address both nutritional deficiencies and environmental challenges.

Future Prospects

The provision of nutritious diets is a challenge due to the rapid growth of the world's population and increasing food demands. Producing food that meets dietary requirements and will be available in the future is crucial for sustainable development. Addressing emerging global food security concerns and accommodating the expanding population, the importance of plant-based proteins continues to grow. Lentils and beans, as pulses, are considered dominant and primary sources of plant-based protein in human diets, playing a vital role in providing essential nutrients and health benefits, and tackling these challenges. Moreover, pulses are essential for improving protein malnutrition in underdeveloped and developing countries.

However, pulses have lower genetic yield potential compared to cereals, which makes them less profitable for farmers and less affordable for consumers. The Green Revolution prioritized cereals and had minimal impact on the pulses sector. As a result, there is a need to address the yield limitations in pulses. It is important to develop efficient protocols for pulse transformation in order to utilize new breeding techniques such as genome editing, taking into account the diverse species of pulses available. The increase in pulse production and consumption is currently crucial. This will not only provide affordable nutrition to the masses but also promote environmental sustainability. While there has been extensive research on the nutritional and health potential of soybeans, the study of other legumes is limited. Therefore, it is essential to conduct comprehensive studies to explore the nutritional potential and health benefits of various legumes.

Conclusion

Meeting the nutritional needs of an expanding global population amid rising food requirements poses significant challenges. Legume-based diets have the potential to ensure food and nutritional security for the rapidly growing global population. Protein-energy malnutrition is a major nutrition-related issue affecting preschool children and lactating women, primarily due to the high cost of animal-based protein and reliance on staple cereal diets. Legumes are considered the most important plant-based protein source in many parts of the world. Legumes are the rich source of proteins, carbohydrates, soluble and insoluble dietary fibers, minerals and vitamins. Due to their high nutritional value, legumes are excellent for combating protein-energy malnutrition and micronutrient deficiencies, especially when combined with grains. Besides their nutritional benefits, legume cultivation is also economically and environmentally sustainable.

References

Colorado State University Extension. (2020). Food smart Colorado. Nutrition, food safety, and health resources from Colorado State University Extension.

Crujeiras, A. B., Parra, D., Abete, I., & Martinez, J. A. (2007). A hypocaloric diet enriched in legumes specifically mitigates lipid peroxidation in obese subjects. Free Radical Res., 41, 498–506.

FAO. (2024). The State of Food Security and Nutrition in the World 2024 – Financing to end hunger, food insecurity and malnutrition in all its forms. Rome.

Jiddere, G., & Filli, K.B. (2015). The effect of feed moisture and barrel temperature on the essential amino acids profile of sorghum malt and bambara groundnut based extrudates. J. Food Process. Technol., 6(5), 448.

Joshi, P.K., & Rao, P.P. (2017). Global pulses scenario: status and outlook. Ann. N. Y. Acad. Sci., 1392(1), 6–17.

McCrory, M. A., Hamaker, B. R., Lovejoy, J. C., & Eichelsdoerfer, P. E. (2010). Pulse consumption, satiety, and weight management. Adv. Nutr, 1, 17-30.

Mubaiwa, J., Fogliano, V., Chidewe. C., & Linnemann, A. R. (2017). Hard-to-Cook phenomenon in bambara groundnut (Vigna subterranea (L.) Verdc.) Processing: Options to improve its role in providing food security. Food Rev. Int., 33(2), 167- 194.

Nedumaran, S., Abinaya, P., Jyosthnaa, P., Shraavya, B., Rao, P., & Bantilan, C. (2015). Grain Legumes Production, Consumption and Trade Trends in Developing Countries. Working Paper Series No 60. ICRISAT Research Program, Markets, Institutions and Policies. Patancheru 502 324, Telangana, India: International Crops Research Institute for the Semi-Arid Tropics. 64 p.

NFA. (2017). The Swedish food composition database. Swedish National Food Agency, Uppsala.

Nieuwenhuis, R., & Nieuwelink, J. (2005). Cultivation of Soya and Other Legumes. In: Nieuwenhuis, R. and M. Voogd (Eds.), 2nd Edn., by Digigrafi, Wageningen, the Netherlands, p 7.

Pittaway, J. K., Ahuja, K. D. K., Robertson, I. K., & Ball, M. J. (2007). Effects of a controlled diet supplemented with chickpea on serum lipids, glucose tolerance, satiety and bowel function. J. Am. College Nutr., 26, 34–340.

Ranjan, S., Sow, S., Ghosh, M., Padhan, S. R., Kumar, S., Gitari, H., et al. (2023). Nutraceutical properties and secondary metabolites of quinoa (Chenopodium quinoa Willd.): a review. International Journal of Food Properties, 26(2), 3477-3491.

Rawal, V., & Navarro, D.K. (2019). The Global Economy of Pulses. FAO, Rome, p. 190.

Sharif, H.R., Williams, P.A., Sharif, M.K., Abbas, S., Majeed, H., Masamba, K.G., Safdar, W. & Zhong, F. (2018) Current progress in the utilization of native and modified legume proteins as emulsifiers and encapsulants—a review. Food Hydrocoll, 76, 2–16.

Sow, S., & Ranjan, S. (2021). A Review on agronomic biofortification for improving food and nutritional security. Research Biotica, 3(2), 139-44.

Suri, D.J., Tano-Debrah, K., & Ghosh, S.A. (2014). Optimization of the nutrient content and protein quality of cereal—legume blends for use as complementary foods in Ghana. Food Nutr. Bull., 35(3), 372–381.

Tripathi, M.K. (2022). Current affairs grain legumes: crops for novel food and nutritional security. Journal of Food Legumes, 35(2), 79-80.

16

Weed Management in Grain Legumes

B.V Asewar[1*], S.U Pawar[2], Apeksha Thombre[2] and Santosh Thombre[2]

[1]*Director of Extension, Vasantrao Naik Marathwada Krishi Vidyapeeth, Parbhani Maharashtra, India*

[2]*Department of Agronomy, Vasantrao Naik Marathwada Krishi Vidyapeeth Parbhani Maharashtra, India*

**Corresponding author's email: bvasewar@vnmkv.org.in*

Abstract

India shares one fourth of global annual grain legume production and is also largest consumer of it. Weeds are the most severe and widespread biological constraint on agricultural production systems and cause damage to cropped and non-cropped lands. They reduce crop yield and degrade the quality of the produce, besides raising the cost of production. The grain legumes are commonly grown under rainfed/dry land conditions and hence do not receive the best management practices that are required for maximization of crop productivity. Among various production inputs, weed management was found to be the most important one contributing about 31% in pigeon pea, 110% in urdbean and 60 % in mungbean towards total productivity. Effective weed management in grain legumes is essential for optimizing crop yield and quality. This process involves a multi-faceted approach that integrates cultural, mechanical, and chemical methods to address the challenges posed by diverse weed species and varying environmental conditions. Traditional single-method approaches often fall short due to the complexity and variability of weed problems like development of resistance in weeds. Therefore, an integrated weed management approach becomes effective, combining timely interventions with adaptive strategies to manage weed competition throughout the growing season. This approach not only enhances crop competitiveness and yield but also offers economic and environmental benefits by reducing reliance on broad-spectrum herbicides and promoting sustainable practices. Continuous monitoring and adjustment of weed management strategies are crucial to effectively tackle evolving weed populations and changing conditions, ensuring long-term success in grain legume cultivation.

Keywords: *Broad-spectrum herbicides, Crop productivity, Crop weed competition, Grain legume, Integrated weed management*

Introduction

The total world acreage under pulses is about 94.14 (Mha) with production of 89.74 (Mt) at 953 kg/ha yields level. India is the largest pulses producing country in the world. It ranks first in area and production with 37 per cent and 29 per cent contribution respectively. Thus, India is the largest producer, processor, importer and consumer of pulses in the world with more than 12 major and minor pulses cultivated over 29-30 million ha. The country witnessed production of 26.05 million tonnes of pulses during 2022-23 (Dixit *et al.*, 2024).

Weeds are one of the major biological constraints that limit the productivity. If quantitatively expressed the loss due to weeds averages around 31.5% (22.7% in rabi and 36.5% in kharif). Weed management in grain legumes is a critical aspect of agricultural practices, essential for optimizing crop yield and quality (Dayal *et al.*, 2023). Weeds are one of the major deterrents that can affect grain legume production systems and in sustaining the crop productivity due to their slow initial growth and short stature. Weeds compete with grain legumes for nutrients, soil moisture, solar radiation and space; and reduce the yield and quality of grains. Besides, they also act as alternate host for insect-pests and disease-causing organisms. Grain legumes, such as pigeon pea, soybean, chickpea, green gram, black gram and lentil are particularly sensitive to weed competition. Weed problems vary in different crops, seasons, agro-ecological conditions and management practices. Effective weed management in grain legumes involves a combination of cultural, mechanical, biological and chemical control methods. Cultural practices, such as crop rotation, cover cropping and proper planting densities help to reduce the weed seed bank and enhance crop competitiveness. Mechanical methods, including tillage and hoeing, physically remove weeds or disrupt their growth. Biological control, through the use of natural weed predators or competitive plant species, offers an eco-friendly approach. Chemical control, using herbicides, remains a common method, but it must be carefully managed to avoid herbicide resistance and environmental impact. Integrated weed management (IWM) strategies are increasingly emphasized in modern agriculture, combining multiple approaches to achieve sustainable and effective weed control. The holistic approach aims to minimize the reliance on herbicides, preserve soil health and protect biodiversity, ultimately ensuring the long term productivity and sustainability of grain legume farming systems.

In general, the yield loss due to weeds is almost always caused by a group of different weed species, and these species can differ considerably in competitive ability (Weaver and Ivany, 1998). Estimating yield loss due to a single weed species is challenging, so it is typically assessed as the total impact of all weed species combined. In India, Bhan *et al.* (1999) estimated and suggested that weeds in India reduce crop yields by 31.5% (22.7% in winter and 36.5% in summer and *Kharif* seasons).

The grain legumes are commonly grown under rainfed/dry land conditions and hence do not receive the best management practices that are required for maximization of crop productivity. Ali and Lal (1989) reported that among various production inputs, weed management was found to be the most important one contributing about 31% in pigeon pea, 110% in urdbean and 60 % in mungbean towards total productivity. In chickpea, weed management contributed the most followed by fertilizer use and insect pest and disease control. Liebman *et al.* (2001) provided a sound basis for the consideration of better understanding the ecological aspects of managing weeds in agro-ecosystems. They identified three primary objectives for weed management. The first is that weed density should be reduced to tolerable levels. However, this concept argues that waiting until weed populations reach an economic injury level makes us more dependent on the use of herbicides. The second objective is that the amount of damage a given density of weeds inflicts on an associated crop should be reduced. The third objective is to shift the composition of weed communities towards less aggressive, easier-to-manage species (Table 1). The competitiveness of different weed species can differ in how much damage they inflict on different crops. Generally, broadleaf weeds are more competitive with broadleaf crops than with grass species (Everman *et al.*, 2008), although there certainly are exceptions (Harker *et al.*, 2007).

Table 16.1: Potential yield loss due to weeds in different major legume crops of India [Rao *et al.* 2014]

Pulse crop	**Crop Yield losses (%)**
Chickpea	10-50
Soybean	10-100
Pea	10-50
Greengram	10-45
Lentil	30-35
Pigeonpea	20-30
Horsegram	30

Weed Flora Observed in Major Grain Legumes

The Crops which are grown during rainy season or under irrigated conditions are more heavily infested with weeds and experience heavy losses in crop yields if weed management is not adequately done. The intensity of weed problem varies with the agro-ecological conditions and the level of management. Carpet weed (*Trianthema portulacastrum*) has been observed as most dominating and aggressive weed occurring in rainy season/summer grain legumes in semi-arid tract of the country. This weed emerges in a number of flushes and produces a large number of seeds. *Celosia argentia, Digera arvensis* and *Commelina benghalensis* are the other major weeds found in grain legumes. As pigeonpea cultivated during the rainy/*Kharif* season, its slow initial growth and wider planting spacing, weed infestation in pigeonpea is particularly severe during the early growth stages compared to other grain legumes. Weeds such as *Commelina benghalensis, Eleusine indica, Setaria viridi, Echinocloa colona/ crusgalli, Brachiaria sp, Panicum sp, Dactyloctenium aegyptium, Digitaria sanguinalis, Digeria arvensis, Euphorbia hirta, Phyllanthus niruri, Tribulus terrestris, Convolvulus arvensis, etc.* has been observed during pigeonpea growing season (Kaur *et al.*, 2015).

Weeds have long been thought to be one of the most important factors contributing to low soybean yield. Weed competition causes yield loss in soybean, ranging from 0-80% depending on the type of weeds, their strength and the time of crop weed competition. (Yaduraju, 2002). During the *kharif* season, a huge number of weed species infest the crop, resulting in lower yields. As a result, one of the key barriers to enhancing soybean output is significant weed competition. Because soybean is a rainy season crop, it is infested with a wide variety of weed flora in various flushes that compete with the crop plants for nutrients, light, and moisture, in addition to their allelopathic effects. The weedflora in soybean included *Trianthema monogyna* L, *Amaranthus viridis* L., *Phyllanthus niruri* L, *Digera arvensis* Forsk, *Cynodon dactylon* (L) Pers, *Echinochloa colonum* (L) weed populations observed in soybean crop are Amaranthus spinosus, Brachiaria reptans, *Ipomoea sp., Xanthium strumarium, Chenopodium abum, Amaranthus sp., Echinochloa crusgalli* and *Euphorbia geniculate.*

In Chickpea among monocots *Brachiria mutica, Echinochloa crusgalli* L. and *Cynodon dactylon* L.; *Cyperus rotundus* among sedges and *Amaranthus viridis* L., *Digera arvensis, Physalis minima, Euphorbia hirta, Parthenium hysterophorus* and *Alternanthera sessilis* among dicots were the prominent weeds at Maharashtra (Gore *et al.,* 2018). *Dactyloctenium aegyptium, Echinochloa colona, Brachiaria* sp., *Cyperus rotundus, Commelina diffusa,*

Amaranthus viridis, Digeria arvensis, Parthenium hysterophorus and *Phyllanthus niruri* are the major weeds observed in greengram (Khairnar *et al.* 2014). The major weeds observed in blackgram are *Cynodondactylon, Dactyloctenium aegyptium, Digitaria sanguinalis, Panicum maximum, Eleusine indica, Cyperus rotundus, Commelina bengalensis, Ageratum conyzoides, Euphorbia hirta, Tribulus terrestris, Trianthema monogynya, Fimbristylis penera, Digera arvensis, Cleome viscosa, Celosia argentia,* etc (Adhikary, 2018).

Critical Weed Competition and Losses Due to Weed Infestation

Weeds one of the greatest biological restrictions to crop output compete with crops for natural and applied resources, as well as reducing agricultural productivity and causing harvesting difficulties (Rao *et al.*, 2015). They also serve as hosts for pests and pathogens. The critical period of weed control is the period of crop growth when the crop must be kept weed-free to prevent yield loss due to weed interference (Van Acker *et al.*, 1993) or critical period of crop-weed competition period is the shortest time span in the ontogeny of crop when weeding results in highest economic returns. Crop Weed Competition is a struggle between crop and weed for a limited resource that is essential for growth. Water, nutrient, light and space are the major factors for which usually competition occurs. Competition between crop plants and weeds is most severe when they have similar vegetative habit and common demand for available growth factors. Naveed *et al.* (2024) graphically depicted data on pest induced yield losses in pulses in Fig. 1.

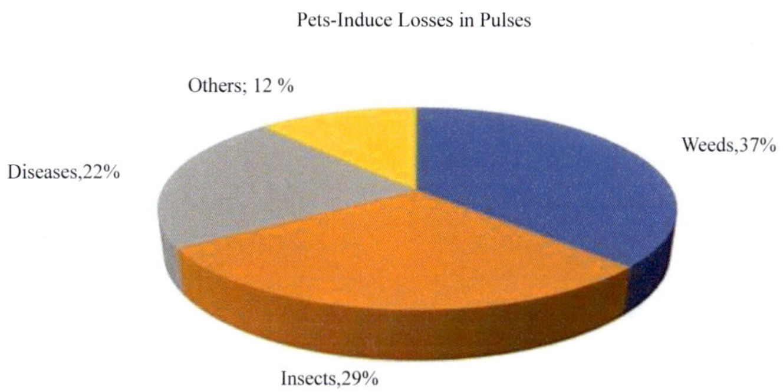

Fig. 16.1: Pest induced yield losses in pulses (Naveed *et al.,* 2024)

Reduction of crop yield has a direct correlation with weed competition. Weeds compete for water, light, nutrients and space. Weeds compete for water in dry land and for nutrients in irrigated crops. It leads reduction in crop yields and production efficiency and erosion of crop quality (Table 2).

The losses due to weeds depends on:

- Type of weed
- Severity of Infestation
- Duration of infestation
- Competitive ability of the crop plants with weeds
- Climatic conditions which affect the growth of the crop and the weed

Weed infestation in pigeonpea is severe at the initial period during first 6-8 weeks, when the crop requires to be kept free from weeds (Kaur *et al.*, 2015). If not carefully managed, weeds can rapidly overtake the crop, leading to yield reductions of 55-60% (Kandasamy, 1999). The critical crop weed competition period in soybean varies from 15–45 days after sowing (DAS) depending upon location and prevailing environmental conditions. Weeds cause a 26–84% reduction in the yield depending upon the types and intensity of weeds, besides impairing the quality of the produce (Kumar *et al.*, 2021). Weeds that persist in soybean fields from three to eight weeks after emergence pose the greatest risk to reducing soybean yields. Weeds that appear six to eight weeks after soybean emergence in fields that have been kept weed-free up to that point are unlikely to significantly impact yield or result in a negative economic outcome relative to the cost of in-season control. However, these later-emerging weeds can still cause harvest issues, lower crop quality, or contribute to an increase in weed seed production. Yield losses caused by weeds in chickpeas have been estimated at 40 to 87 per cent in India, 41 to 42 per cent in the former Union of Soviet Socialist Republics (USSR). However, studies at Indian Institute of Pulses Research, yield losses due to weed in chickpea varied between 29-70 percent over the years (Anonymous, 2009). The critical period of weed competition in green gram has been reported to be between 3 and 6 weeks after planting (Utomo *et al.*, 1988). Unchecked weeds were reported to cause a considerable reduction in black gram seed yield (upto 46-53%) in summer (Bhandari *et al.*, 2004) and *Kharif* season (43.2-64.1 %) (Rathi *et al.*, 2004). Following table shows yield losses due to weed infestation in major legume crops reported by different authors.

Table 16.2: Crop yield loss due to weeds in different major legume crops of India

Crop	Yield loss (%)	Reference
Pigeonpea	31.0-52.8	Singh and Sekhon (2013)
Chickpea	77.8	Patel *et al.* (2006)
Urdbean	43.3	Choudhary *et al.* (2012)
Mungbean	38.6	Kumar *et al.*(2016)
Lentil	37.6	Yadav *et al.*(2013)
Fieldpea	50.0	Singh*et al.* (2016)
Rajma	49.5	Kumar*et al.* (2014)
Lytharus	46.10	Pramanik *et al.*(2014)
Mothbean	30-40	Bathar*et al.* (2005)

Methods of Weed Control In Grain Legumes

Weed control is the process of limiting infestation of the weed plant so that crops can be grown profitably, where as weed management includes prevention, eradication and control by regulated use, restricting invasion, suppression of growth, prevention of seed production and complete destruction. For successful crop cultivation, it's essential to follow certain basic prevention measures like maintaining a clean cultivation process, using clean seeds, keeping the seed bed and irrigation channels weed-free, applying well-decomposed organic manures, and ensuring tools and farm machinery are clean. Thus weed control is one of the aspects of weed management. Weed control includes

1. Cultural measures for weed control
2. Biological. Measures and
3. Chemical weed control

Cultural Measures for Weed Control

Prior to the introduction of chemical herbicides, cultural practices were crucial in enhancing crop competitiveness against weeds at minimal cost. This eco-friendly approach aimed to reduce the introduction, establishment, and spread of weeds in new areas. Some cultural practices such as choice of crop species, crop cultivars, planting density, crop geometry, inter cropping, crop rotation, time of sowing, crop rotation, fertilizers and irrigation practices have profound effect on weed suppression. It has been demonstrated that a higher density of legume as cover crops also increases competitive interactions and resource consumption between the main crop species. This in turn provides an effective weed reduction (Elsalahy, 2019). With greater coverage of the soil surface with CCs at an early stage of their growth a more effective suppression of weeds was observed, especially at a high sowing density of species.

In wide spaced pulses such as pigeonpea, intercropping of sorghum, maize, greengram, blackgram and groundnut is a common practice, which besides covering risk will reduce weed infestation. In pigeonpea cultivation and pigeonpea-based cropping systems, crop rotation and intercropping are crucial for effective weed suppression.

Cover crops namely ryegrass, oats, rye, pigeon pea and velvet bean are used in soybean to reduce the weed population, weed biomass, weed seed bank and improve soil physical, chemical and biological status. Although, the time of application of fertilizers is found to affect weed population. In chickpea, increasing crop densities by adjusting seed rates and row spacing effectively reduces the areas available to weeds and inhibits their growth. In greengram, cultural practices such as, use of mulch, tillage operation, maintaining row spacing, use of different genotypes, crop rotation are followed.

Table 16.3: Weed-Smothering Efficiency (WSE) of different pulse-based intercropping systems [Ali, 1988]

Intercropping systems	WSE (%)
Pigeon pea + Black gram	32.8
Pigeon pea + Green gram	31
Pigeon pea + Cowpea	39.1
Pigeon pea + Sesame	36.6
Pigeon pea + Pearl millet	50.8
Black gram + Maize	17.3
Pigeon pea + Maize	16.4

Mechanical Weed Control Measures

Controlling weeds before they reach their reproductive stage is crucial. Beyond cultural practices, weeds can be managed physical methods such as hand hoeing, hand pulling, tillage, flooding, desiccation, burning and mulching. Hand hoeing and manual weeding are two most common practices performed for weed management in almost all grain legumes. One or two hand weeding at critical crop-weed stage provide satisfactory control of weeds in most of the crops.

These methods are efficient, cheaper, safer, to crop and no harmful effect to crop and user. Oldest, effective and economical method. No special skill is required in adopting physical methods.

In pigeonpea, two mechanical weeding, one at 25–30 DAS and another at 45–50 DAS have been found to be adequate. In green gram and black gram, two hand weeding at 25 and 45 DAS are found effective in controlling weeds. In winter season pulses, *viz.* chickpea, lentil and French bean, two weeding at 30 and 60 days after sowing provide effective control of all the weeds. In soybean, application of hoeing and hand weeding twice significantly reduced the weed count and weed dry weight as compared to the weedy checks (Dhaker *et al.,* 2015). Numan *et al.* (2017) found two times hand weeding is the most economical and beneficial method for controlling weeds in chickpea and increasing yield and yield components due to maximum nutrient availability for the crop. One hand weeding/hoeing at 30-35 days after sowing or application of a pre-emergence herbicide like Pendimethalin @ 0.75 to 1 kg a.i./ha in 500-600 liters of water immediately after sowing helps to keep the losses by weeds below ETL (Directorate of Pulses Development, https://dpd.gov.in)

Chemical Weed Control

Herbicides are the most successful weed control technology ever developed as they are selective, cost effective, fairly easy to apply, have persistence that can be managed, and offer flexibility in application time. Herbicides have come as a big boon to farmers in areas where the labour supply is limited and wages are high.

Some of the unintended negative impacts of herbicide use are persistence in soil, pollution of ground water, toxic residues in food (contamination), feed and fodder, adverse effect on nontarget organisms and development of resistance in weeds. In India, the use of herbicides in pulse cultivation is minimal, primarily due to the limited purchasing power and risk tolerance of dryland farmers. However, experimental evidence and on-farm demonstrations have demonstrated that herbicides like alachlor, pendimethalin, and fluchloralin are highly effective in controlling seasonal weed flora (Roy *et al*., 2023).

In a pigeonpea/sorghum intercropping system, the pre-emergence application of pendimethalin at 1.5 kg/ha has been shown to be highly effective in controlling weeds. For pigeonpea combined with short-duration legumes, fluchloralin at 0.5 to 0.75 kg/ha and alachlor at 2 kg/ha have demonstrated effectiveness in managing seasonal weeds and boosting productivity (Venkateswarlu and Ahlawat 1986). In pigeonpea, pre-emergence applications of pendimethalin 1.25 kg/ha was found most effective with 21.4% higher grain yield (Kaur *et al*., 2015). Herbicides are highly effective for managing weeds in soybean crops. Since soybeans are grown during the rainy season, relying solely on mechanical methods for weed control can be risky due to continuous rainfall, which often prevents their use. In such cases, using herbicides offers a reliable solution for effective weed control.

Dewangan *et al.* (2016) revealed that with the use of tank-mix pre-emergence application of Oxyfluorfen + Metribuzin @125 + 350 g ha^{-1} there was a notable increase in the yield (73.10%), over the control. While Metribuzin @250 g ha^{-1} at 20 DAS and Oxyfluorfen @125 g ha^{-1} at 12 DAS showed phytotoxic effects on chickpea plant. S-metolachlor, imazethapyr, linuron (low rate) and s-metolachlor + imazethapyr decreased kidney bean yield 41, 29%, 35% and 24% respectively compared to the weed free control treatment, however, linuron (high rate), s-metolachlor + linuron and s-metolachlor + imazethapyr + linuron applied PRE treatments provided comparable yield to the weed free control treatment (Soltani *et al* 2014). In greengram, sequential application of pendimethalin as PE followed by imazethapyr as PoE can also be done for controlling weeds (Singh and Singh, 2020).

Table 16.4: Herbicides approved for use in major grain legumes

Crops	Approved Herbicide name	Weed species	Dosage /ha		Dilution in Water (Litres)
			a.i. (gm/ Kg)	Formulation in (gm/ ml /Kg/ ltr)	
Soybean	Chlorimuron Ethyl 25% WP + Surfactant	*Cyperus rotundus Commelina benghalensis Celosia argentea Digera arvensis Cucumis trigonus Cyprus iria, Parthenium hysterophorus, Acalypha indica, Phyllanthus niruri, Trianthema portulacashurm, Caesulia auxillaris*	9 gm (3-15DAS)	36 gm.	300 ltrs. + non ionic surfactant 0.2 % (Iso- octyl phenoxyl - poloxetha nol 12.5 %)
	Diclosulam 84% WDG	*Cyperus spp, Commilena benghalensis, Euphorbia geniculata, Digera arvensis, Acylipha spp, Echinochlo colona*	22-26 g	26.2-30.9 Time of application 0-3 DAS	500
	Pendimethalin 30% EC	*Echinochloa spp., Euphorbia spp., Amarnanthus viridis, Portulaca oleracea, Trianthema spp., Eleusine indica*	0.75-1.0 kg	2.5-3.3 ltrs.	500-700
	Imazethapyr 10% SL	*Cyperus difformis Echinochloa colonum E. crusgalli Euphorbia hirta Croton sperrsifeorus, Digera arvensis, Commelina Benghalensis*	100 g	1.0 Ltr.	500
	Fluazifop-p-butyl 11.1% w/w + Fomesafen 11.1% w/w SL	*Echinochloa colona Digitaria sp Eleusine indica Dactyloctenium aegyptium Brachiaria reptans Commelina benghalensis Digera arvensis Trianthema sp. Phyllanthus niruri Aclypha indica Dinebra arbica*	250	1000	500
	Fomesafen 12 % + Quizalofop ethyl 3% w/w SC	*Echinochloa colona, Eleusine indica, Doctyloctenium aegyptium, Digitaria sanguinolis, Cyperus rotundus*	180+45	1500	500

	Imazethapyr 35% + Imazamox 35% WG	*Echinochloa Colonum, Dinebra Arabica, Digitaria sanguinalis, Brachiaria mutica Commelina benghalensis Euphorbia hirta*	70 g a.i/ ha + MSO Adjuvant @ 2 ml/l of water	100 g MSO Aadjuvant @ 2ml/l of water	375-500
Red gram	Pendimethalin 30% EC	*Digitaria sanguinalis Digera arvensis Amaranthus sp. Euphorbia hirta Trianthema sp. Cyperus sp. Eragrostis sp.*	0.7 – 1.00	2.5 – 3.33	500
	Imazethapyr 35% + Imazamox 35% WG	*Euphorbia spp., Amaranthus viridis.*	70 g a.i./ ha + MSO adjuvant @ 2m/l of water	100 g/ha + MSO adjuvant @ 2m/l of water	375-500
Green gram	Imazethapyr 10% SL + Surfactant	*Amaranthus viridis Trianthema portulacastrum Euphorbia hirta*	75 ga.i.lha+ MSO adjuvant @2 ml/litre of water	750 ml+MSO adjuvant @ 2ml/ha of water.	375
Blackgram	Fenoxaprop-p-ethyl 9.3% w/w EC (9% w/v)	*Echinochloa crusgalli, Echinochloa colona Digitaria sp. Dactylocteneum Aegyptium*	56.25-67.5 g.	625-750ml (15-20 DAS)	375-500

[Source: Major Uses of Pesticides (Registered under the Insecticides Act, 1968) UPTO - 31/03/2024]

Integrated Weed Management Practices

An integrated weed management approach involving a combination of both chemical and agronomic manipulations resulted in improved crop geometry and density which resulted in a better crop competitive ability against weeds. Dhonde *et al.* (2009) found that the highest seed yield of pigeonpea (2.30 t/ha) was achieved with a weed-free treatment, followed by an Integrated Weed Management (IWM) approach involving pendimethalin at 1.0 kg/ha combined with hand weeding at 45 days after sowing. Similarly, Talnikar *et al.* (2008) reported that the pre-emergence application of alachlor at 2 kg/ha, in conjunction with hand weeding and hoeing at 6 weeks after sowing, was the most effective and economical method for controlling weeds and increasing grain yield in a pigeonpea and soybean intercropping system.

Chandraker and Paikra (2015) concluded that, pre-emergence herbicide Pendimethalin (1000 g/ha) followed by hand hoeing at 20 DAS gives better results in controlling weeds in soybean. Singh *et al.* (2003) found application of pendimethalin and isoproturon pre- or post-emergence and application of pendimethalin pre-emergence in combination with hoeing 45 DAS were effective in chickpea. Integration of herbicides with HW generally provides efficient weed control without any negative influence on symbiosis, growth, yield and nutrient uptake of greengram (Singh and Singh, 2020).

Table 16.5: Most economical IWM methods for managing weeds in certain crops of India

Crop	IWM	Reference
Blackgram	(i) Pendimethalin at 0.75 kg/ha fb HW 45 DAS (ii) Pendimethalin at 0.50 kg/ha fb HW 60 DAS (iii) Trifluralin (PE) at 0.50 kg/ha fb HW	i) Kumar *et al.* (2006) (ii) Rathi *et al.* (2004) (iii) Sardana *et al.* (2006)
Cowpea	Pendimethalin 0.75 kg/ha fb HW 35 DAS	Mathew *et al.* (1995)
Dwarf pea	Sowing at 20 cm apart with two HW fb pendimethalin at 1 kg/ha	Tewari *et al.* (2003)
Pigeonpea/ Groungnut intercrop	Pendimethalin (1.0 kg/ha) or fluchloralin (1.0 kg/ha) each fb	Vijaykumar *et al.* (1995)
Pigeonpea/pearl millet intercrop	Pendimethalin at 1.50 kg/ha+HW 40 DAS	Shinde *et al.* (2003)
Soybean	(i) Butachlor (Pre-em) 1.5 kg/ha fb HW 30 DAS (ii) Rows spacing of 22.5 cm with alachlor at 1 kg/ha	(i) Chandrakar and Urkurkar (1993) (ii) Shekara and Nanjappa (1993)

DAS–Days after seeding, DAT–Days after transplanting, DAP–Days after planting, HW–Hand weeding, fb–followed by, PE–Pre-emergence.

Conclusion

Weeds compete intensely with crops for essential growth resources. In addition to decreasing grain yield, weeds can lower produce quality and serve as alternate hosts for disease-causing organisms and insect pests. Therefore, it is crucial to integrate various weed management practices effectively during the critical period of crop-weed competition. Weed management in food legume crops presents several challenges, including the intermittent emergence of weeds, unpredictable rainfall, unsuitable soil conditions, and a lack of timely labor and broad-spectrum herbicides. Considering the complexity of the weed problem, no single method whether manual, mechanical, or chemical can offer comprehensive, season-long control under all circumstances. Considering the diversity of weed problem, an integrated weed management system, as part of a broader integrated crop management approach, would be an effective, cost-efficient, and environmentally friendly method for managing weeds in grain legumes. Continuous evaluation and adjustment of weed management practices are necessary to address evolving weed populations and changing environmental conditions, ensuring long-term effectiveness and sustainability.

References

Adhikary, P. (2018). Weed management in blackgram. Indian Journal of Weed Science, 50(4), 369–372.

Ali M. (1988). Weed suppressing ability and productivity of short duration legumes intercroppped with pigeonpea under rainfed conditions. International Journal of Pest Management, 34(4), 384-387

Ali, M., & Lal, S. (1989). Priority inputs in Pulse Production. Fertilizer News, 34, 17-21.

Anonymous, (2009). Twenty-five years of Pulses research at IIPR Kanpur. 1984-2009. Indian Institute of Pulses Research, Kanpur.

Bhan, V. M., Sushilkumar., & Raghuwanshi, M. S. (1999). Weed Management in India. Indian J. Plant Prot, 17, 171- 202.

Bhandari, V., Singh, B., Randhawa, J. S., & Singh, J. (2004). Relative efficacy and economics of integrated weed management in blackgram under semi-humid climate of Punjab. Indian Journal of Weed Science, 36, 276–277.

Chandraker, A. K., & Paikra, P. R. (2015). Effect of integrated weed management on weed dynamics of soybean (Glycine max L. Merrill) under Chhattisgarh plain. Indian Journal of Agricultural Research, 49, 53–58.

Dayal, P., Kumar, A., Pal, R. K., Sow, S., Ranjan, S., Tyagi, S., & Kumar, P. (2023). Weed dynamics, productivity and profitability of wheat as influenced by tillage and weed management practices in Eastern Indo-Gangetic Plains. Indian Journal of Weed Science, 55(4), 453-460.

Dewangan, M., Singh, A. P., Chowdhury, T., Kumar, D., & Kumar, B. (2016). Management of complex weed flora in chickpea. Indian Journal of Weed Science, 48(1), 79-82.

Dhaker, S. C., Mundra, S. L., Dhaker, R. C., & Sumeriya, H. K. (2015). Effect of weed management and sulphur on nutrient content and uptake by weeds and soybean. Legume Research, 38(3), 411–14.

Dhonde, M. B., Kate, S. R., Pandure, B. S., & Tambe, A. D. (2009). Integrated weed management in pigeonpea. Indian Journal of Weed Science, 41(1&2), 102-105.

Dixit, G. P., Srivastava, A. K., & Ali, H. (2024). Scenario of pulses production in India. Indian Farming, 74(2), 03-06.

Elsalahy, H., Döring, T., Bellingrath-Kimura, S., & Arends, D. (2019). Weed Suppression in Only-Legume Cover Crop Mixtures. Agronomy, 9, 648.

Everman, W. J., Burke, I. C., Clewis, S. B., Thomas, W. E., & Wilcut, J. W. (2008), Critical period of grass vs broadleaf weed interference in peanut. Weed Technol., 22 (1), 68–73.

Gore, A. K., Chavan, A. S., Gokhale, D. N., & Thombre, K. M. (2018). Evaluation of New Herbicides on Weed Flora and Productivity of Chickpea (Cicer arietinum L.). International Journal of Current Microbiology and Applied Sciences, 7(5), 3682-3687.

Harker, K. N., Blackshaw, R. E. & Clayton, G. W. (2007). Wild oat (Avena fatua) vs. redstem filaree (Erodium cicutarium) interference in dry pea. Weed Technol., 21 (1), 235–40.

Hauggaard-Nielsen, H., Andersen, M.K., Jørnsgaard, B., & Jensen, E.S. (2006). Density and relative frequency effects on competitive interactions and resource use in pea-barley intercrops. Field Crop. Res., 95, 256–267.

Kandasamy O. S. (1999). Effect of herbicides with and without manual weeding on weeds and yield of rainfed pigeonpea (Cajanus cajan L. Millsp.). Legume Research, 22(3), 172-176.

Kaur, R., Raj, R., Das, T. K., Shekhawat, K., Singh R., & Choudhary A. K. (2015). Weed management in pigeonpea-based cropping systems. Indian Journal of Weed Science, 47(3), 267–276.

Khairnar, C.B., Goud, V. V., & Sethi, H. N. (2014). Pre- and postemergence herbicides for weed management in mungbean. Indian Journal of Weed Science, 46, 392–395.

Kumar, S., Rana, S. S., & Ramesh. (2022). Weed management strategies in Soybean (Glycine max) — A review. Indian Journal of Agricultural Sciences, 92(4), 438–44.

Liebman, M., Mohler, C. L., & Staver, C. P. (2001). Ecological Management of Agricultural Weeds, Cambridge, England: Cambridge University Press.

Naveed, M., Asad, M. A., Tariq, M. S., & Aslam, M. (2024). Induced Mutation: A New Paradigm in Pulse Weed Control Strategies, Intech open. doi:10.5772/ INTECHOPEN.1005246

Numan, A., Ali, N., Khan, M. I., Rehman, M. U., & Qadir, A. (2017). Yield and yield parameters of chickpea varieties under different weed population in Thal area of Khyber Pakhtunkhwa. International Journal of Green and Herbal Chemistry, 7(1),119-129.

Rathi, J. P. S., Tewari, A. N., & Kumar, M. (2004). Integrated weed management in blackgram (Vigna mungo L.). Indian Journal of Weed Science, 36, 218–220.

Roy, D. K., Ranjan, S., & Sow, S. (2023). Productivity and profitability of zero till winter maize as influenced by integrated weed management practices. Indian Journal of Weed Science, 55(3), 264-268.

Singh, A., Vashist, K. K., & Kang, J. S. (2003). Chemical weed control in irrigated desi gram (Cicer arietinum L.). Indian Journal of Weed Science, 35(1 and 2), 136-138.

Singh, R., & Singh, G. (2020). Weed management in greengram: A review. Indian Journal of Weed Science, 52(1), 10–20.

Soltani, N., Nurse, R. E., & Sikkema, P. H. (2014). Weed management in kidney bean with tank mixes of S-metolachlor, imazethapyr and linuron. Agricultural Sciences, 5(7), 46953.

Talnikar, A. S., Kadam, G. L., Karande, D. R., &, Jogdand, P. B. (2008). Integrated weed management in pigeonpea [Cajanus cajan (L.) Millsp.] International Journal of Agricultural Sciences, 4(1), 363-370.

Utomo, I. H. (1988). Critical period of mungbean (Phaseolus radiatus L.) to weed competition. Biotropia, 2, 8–11.

Van Acker RC, Swanton CJ, Weise SF. (1993). The critical period of weed control in soybean (Glycine max (L.) Merill). Weed Sci., 41, 194-200.

Venkateswarlu, U., & Ahlawat, I. P. S. (1986). Studies on weed management in pigeonpea based intercropping systems. Indian Journal of Agronomy, 31, 184-187.

Weaver, S. E., & Ivany, J. A., (1998). Economic thresholds for wild radish, wild oat, hemp-nettle and corn spurry in spring barley. Can. J. Plant Sci., 78, 357–361.

17

Weed Seed Bank: Impacts and Management Strategies for Grain Legumes Production

Pooja A. P.[1*] and Arunjith P.[2]

[1]Agricultural Officer, Agriculture Development and Farmers Welfare Department Government of Kerala

[2]Specialist (Soil Science), Kerala State Land Use Board, Government of Kerala

**Corresponding author's email: appoojaap@gmail.com*

Abstract

Legumes are an excellent group of fertility restorer crops which can be grown either as sole crop or as a component crop in the cropping system. As they are tolerant to drought condition, they can be grown throughout the year. The yield of legume crop is largely depending on pest, diseases and weeds. Among these weeds plays a major role as the weed seed bank inside the soil influence the weed population in the cropping season. Hence it is important to manage the weed seed bank for better productivity of legumes. Though legumes are short duration and fast growing crops, the germination percentage of weed seed bank will determine the final yield of legumes. There are many strategies to which the weed seed bank can be controlled to tolerable level before which the weed seed bank present in the soil has to estimate. Annual weed seed bank can be minimized by killing the weeds before they set seeds, harvest weeds, crop rotation, cover crops, inter cropping and ensiling. Weed seed bank depletion can be done by tillage, tricking the weed seeds, conditioning weed seeds, soil solarisation and herbicide application. Thus integrating various technologies may help in effective controlling of weed seeds to tolerable level thereby enhancing the yield of leguminous crops.

Keywords: *Estimation of weed seed bank, legumes, tillage, yield, weed seed bank*

Introduction

Weed seed bank can be defined as the sum of viable weed seeds and vegetative propagules that are present in the soil and thus contribute to weed pressure in future crop (Menalled, 2013). It is the primary source of new infestation. It includes both the new weed seeds recently shed and the older seeds. Grain legumes differ in their competitive ability with weeds. Some intensively cropped grain legumes are effective in altering weed populations and species for subsequent legume crops (Cardwell, 1987). The weed populations that occur in a legume cropped area and the success of weed management are influenced by weed seed bank characteristics. Changes in seed bank characteristics can be made by changing the management practices which will alter the pattern of disturbance of legumes and thereby yield (Sahoo *et al.*, 2024).

In agronomic situations, the majority of the seeds that germinate during the first few years represent the major threat for yield loss and control costs in grain legumes. While the persistent portion of the seed-bank is important in long-term population dynamics, understanding the short-term dynamics of seed banks will aid in predicting crop yield losses and control costs. The weed seed bank serves as a physical history of the past successes and failures of cropping systems, and knowledge of its content (size and species composition) can help producers both anticipate and ameliorate potential impacts of crop weed competition on grain yield and quality in legumes.

Travelling of Weed Seeds

Weed seeds can reach the soil surface and become part of the soil seed bank through several avenues (Fig. 1). The main source of weed seeds in the seed bank is from local matured weeds that set seed. Agricultural weed seeds can also enter a field by animals, wind, water and human activities, like cultivation and harvesting. How far weed seeds can travel depends on the dispersal process and the weed species.

Components of Weed Seed Bank

Soil seed bank include:

- Seeds - mostly annual plant
- Bud bank-vegetative propagules which include dormant and adventitious buds, stolon, rhizomes and bulbs
- Diaspore-produced by non-flowering plants such as ferns and bryophytes

Generally, seed banks are composed of many species, with a few dominant species comprising 70 to 90% of the total seed bank (Wilson, 1988). These species are the primary pests in agronomic systems because of resistance to control measures and adaptation to the cropping system. A second group of species, comprising 10 to 20% of the seed bank, are generally those adapted to the geographic area but not to current production practices. The final group accounts for a small percentage of the total seed and includes recalcitrant seeds from previous seed banks, newly introduced species, and seeds of the previous leguminous crop (Wilson *et al.*, 1985). This group undergoes constant change due to seed dispersal by humans, other animals, wind, and water.

Table 17.1: Seed production capacity (Walia, 2014)

Weed Species	**Number of seeds per plant**
Amaranthus spp.	196000
Chenopodium album	72000
Brassica nigra	58000
Commelina benghalensis	2450
Cynodon dactylon	170
Cyperus esculentus	820
Datura stramonium	1239000
Eleusine indica	41200
Physalis angulate	22000
Portulaca spp.	193000

Persistence, Longevity and Fate of Weed Seeds under Grain Legumes

Types of Seed Bank: Based on longevity seed bank can be divided into:

- **Transient:** Die or germinate within one year of introduction.
- **Persistent:** Seeds remain viable in the soil year after year.

Weed Seed Persistence

Persistence capacity is the adaptive potential of a weed that enables it to grow in any environment. It is the consequence of a large number of factors like a great number of seeds produced, long term viability, continuous germination, phenotypic and genotypic plasticity.

Seed Longevity

Seed longevity in the soil depends on the interaction of many factors including the intrinsic dormancy of the seed population, the environmental conditions (e.g. light, temperature, moisture) and biological processes (e.g. predation, allelopathy). Understanding how management practices or environmental conditions modify the residence time of viable seeds can help producers to minimize future weed problems. For example, tillage enhances seed longevity as weed seed usually remain viable longer if they are buried. On the other hand, no-till reduces seed persistence by exposing them to predators and pathogens.

Fate of weed seeds in soil

Weed seeds can have numerous fates after they are dispersed into a field (Fig. 17.1). Of the many seeds in the seed bank, very few will actually emerge and produce a plant. Most seeds will die, decompose or be eaten before ever germinating. Of those that do germinate, some will die before a mature plant is produced. Seed predation is typically greatest when weed seeds remain on the surface and there is sufficient residue cover for predators (i.e. no-till). Generalist predators such as common ground beetles or crickets can reduce weed seed emergence by 5 to 15%. Larger animals such as rodents and birds can also consume significant amounts of weed seeds. When buried and not available to predators, attack by pathogens is more common (Roy *et al.*, 2024).

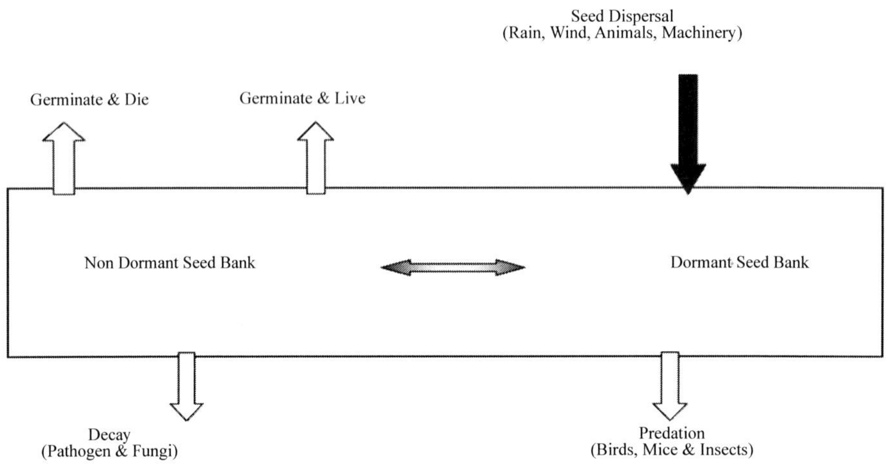

Fig. 17.1: Fate weed seeds in soil (Menalled, 2008)

In practice, managing seed banks is far more complex because it makes difficult in preventing seed production and introduction, there will be persistence of a small percentage of the seed bank and high seed production potential of many weed species. There come the old gardeners saying that "one year's seeding is seven years weeding".

Strategies for Weed Seed Management in Grain Legumes

Various strategies for managing soil seed bank includes:

I. Soil seed bank estimation in grain legumes

Quantifying weed seeds in the soil seed bank enables to determine the potential density, intensity and diversity of weed infestation in legumes cropped area. This information is useful in understanding and predicting the fate of weed seeds in the soil and designing effective, viable and economic weed management programmes. A number of methods have been described for collecting soil samples and extracting weed seeds. These include Augers, metal tubes, plastic liners and vacuum system for collecting soil samples and adopting the techniques of sieving, floatation in high density solutions and centrifugation for extracting and separating weed seeds. These methods asides from time consuming are effective in estimating the densities of only some weed species in the soil.

Methods for weed seed bank analysis

1. **Separation method:** Seeds collected from soil sample are counted. Elutriation is the process for separating particles based on their size, shape and density, using a stream of gas or liquid flowing in a direction usually opposite to the direction of sedimentation.

Elutriator

The elutriator uses water (at pressure 275-345 kPa) to separate weed seeds from clay and silt particles in cores. The principle components of the elutriator include a hooded housing unit containing the spray nozzles, rotating drum, water basin and cylindrical strainers, a rheostat-controlled motor and a sediment trap (Fig. 17.2). The rotating drum made of stainless steel, is mounted to the housing unit. It contains 36 capped strainers (each with 6.5cm diameter, 17.5 cm long and lined with 318 μ stainless steel screens) which allow water to pass through while retaining weed seeds, plant debris, stones and large sand particles. The water basin is filed with 10L of water before drum starts turning. As the drum rotates a continuous stream of water is sprayed into the strainers from six TeeJet 15006 nozzles mounted in the elutriators hood. Excess water and sediment flow into the sediment trap through an overflow outlet in the water basin.

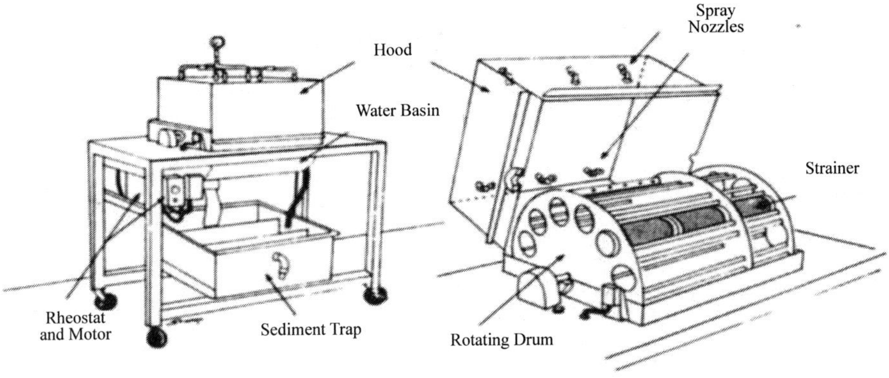

Fig. 17.2: Elutriator and principal components (Wiles *et al.*, 1996)

Using a flushing device the sample contents in each strainer is washes into propyltex fabric bag for drying and storage (Fig. 3). Bags are 12 cm wide and 12 cm long on one side and 16 cm long the other side (to allow a 4cm flap when folded don to prevent loss of the sample contents). The contents are dried overnight and counted and identified using microscope.

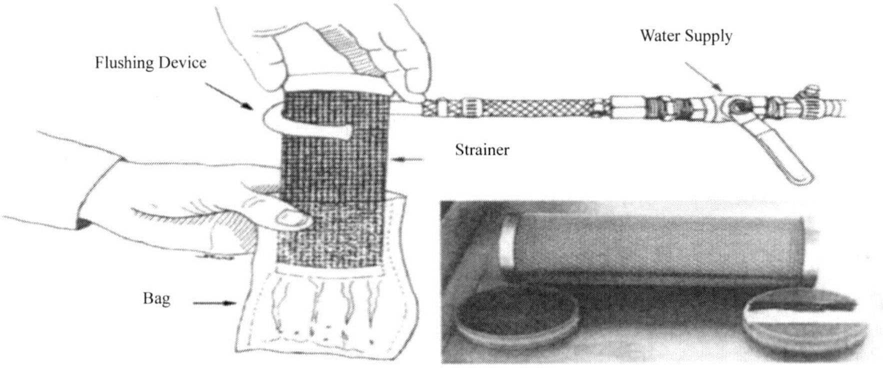

Fig. 17.3: Flushing device for transferring the sample content from a strainer to a propyltex fabric bag and a photograph of a strainer and end caps (Wiles *et al.*, 1996)

Flotation of soil organic matter on a heavy liquid (K_2CO_3-1.56) followed by sieving and hand separation and then check viability by germination can also be used (Kropac, 1966).

2. **Vegetation methods:** Seed bank is assessed via identification and enumeration of seedlings emerged from the soils under controlled conditions.

Weed seed bank estimation (Luschei, 2003)

1. Composite soil samples were collected from plot using a soil auger at a depth of 15 cm.
2. From the composite samples one kg soil was weighed accurately and transferred and spread evenly in plastic trays under net house condition (soil was kept at adequate moisture level).
3. The emerging weeds were counted up to 70 days at fortnightly interval and categorized into sedges, broad-leaved weeds and grasses.
4. The data generated were statistically analyzed using ANOVA.

Here in vegetation method Samples observed under favourable conditions for germination and germinating species are identified and removed periodically.

- **Advantage:** provide the most complete listing of species present in the seed bank.
- **Disadvantage:** Underestimates the seed bank because germination requirements for some species will not be met.

Rapid Method for Enumeration of Viable Seeds (Malone, 1967)

The procedure requires a solution of 10 g of sodium hexametaphosphate (Calgon), 5 g of sodium bicarbonate (baking soda), and 25 g of magnesium sulphate (Epsom salts) dissolved in 200 ml of tap water for each 100 g of soil. The first two chemicals induce the breakdown of soil aggregates and ensure efficient extraction of debris by flotation, accomplished by the magnesium sulphate. The temperature of the solution is of no consequence but must be such that seed viability is not altered. Tap water normally is about 20°C and is sufficient.

Once the soil sample is introduced, agitation can be either by mechanical means or by hand and should proceed for about 2 min. Following this, the floating debris containing the seeds can be (decanted from the solution. A sieve with 0.140-mm openings is sufficient for retaining the seeds. Once the extracted material is on the sieve, it may be rinsed gently with tap water to remove foam and minute soil particles.

Once the organic debris is collected, the seeds can be removed and sorted into species groups. This task is facilitated by air-drying the debris and locating the seeds with the aid of a binocular stereomicroscope. Per cent viability of the seeds can be determined by the tetrazolium chloride test, requiring about 24 hr.

Comparison of Three Methodologies for Efficient Seed Extraction in Studies of Soil Weed Seed Banks (Mesgaran *et al.*, 2007)

They have compared methods viz., sieving, cloth bag and flotation and soil samples of four textures, namely clay, clay loam, loam and sandy loam containing 19%, 27%, 38% and 65.5% sand respectively. Counting time in flotation, cloth bag and sieving methods was 9, 16 and 30 min, respectively. They also reported that species-wise seed recovery was not affected by soil texture in the bag method, suggesting an advantage for this approach as its accuracy is not soil texture dependent.

II. Minimizing annual deposits to the weed seed bank

i) Kill the weeds before they set seeds

ii) Harvest weeds

Concurrent maturity of crops and infesting annual weeds means that commercial crop harvest can result in the "harvesting" and subsequent redistribution to the soil surface of seed of weed species (Walsh and Powles, 2014).

- Harvest weed seed control (HWSC) systems target weed seed during commercial grain harvest operations and act to minimize fresh seed inputs to the seed bank

- These systems exploit two key biological weaknesses of targeted annual weed species: seed retention at maturity and a short-lived seed bank
- HWSC systems, including chaff carts, narrow windrow burning, bale direct, and the Harrington Seed Destructor, target the weed seed bearing chaff material during commercial grain harvest
- The destruction of these weed seeds at or after grain harvest facilitates weed seed bank decline, and when combined with conventional herbicide use, can drive weed populations to very low levels

In *kuttand* area, (rice bowl of Kerala, lies below mean sea level) farmers are practicing a method for managing weedy rice by cutting the panicles of weedy rice before harvesting. This will reduce contamination with harvested produce and development of further seed bank in soil.

iii) **Legume and non-legume rotation:** Crop rotation is also an effective means of managing the weed seed bank especially in leguminous crop. It not only alleviates the problem of weed seed bank, but also helps in maintaining soil fertility and pest and disease infestation. Introducing perennial crops in annual cropping systems tends to deplete the soil seed bank of annual species over time. This method is more effective on weed species which have low levels of longevity such as kochia and many of the grassy weeds like wild oat and green foxtail. Likewise, crop competition is also important for decreasing weed seeds being recruited to the seed bank. Studies near Saskatoon, SK conducted in the late 1970s showed that seed bank populations were greatest in summer fallow (about 1600 seeds m^{-2}) versus wheat stubble (about 500 viable seeds m^{-2}). Weed seed bank additions are high in fallow fields in part due to incomplete weed control by tillage and the absence of a competitive crop.

iv) **Cover crops:** Legumes can serve as cover crops too. Cover crops will reduce soil erosion. Through proper water management and nutrient management, it will help to reduce the weed density. They will act as mulch material.

v) **Inter cropping:** Inter spaces can be effectively utilized for cultivation. Like legumes, they will reduce soil erosion. Through proper water management and nutrient management, it will help to reduce the weed density. They will act as mulch material.

vi) **Ensiling:** Ensiling of forages will reduce weed seed bank replenishment. Use of perennial or annual forages which are harvested prior to seed maturation can be used for ensiling.

III. Depletion of weed seed bank

There are several ways to deplete the weed seed bank. Which include;

i) **Tillage:** Tillage was the main method for managing weeds, until the introduction of herbicides. The degree of soil inversion and depth of tillage, strongly affected the vertical distribution of weed seeds in the soil seed bank. When using a moldboard plow, 37% of the viable weed seed bank was found in the top 5 cm of the soil profile and 74% under no-till. Using a chisel plow resulted in 61% of the seed near the soil surface. Deep buried seeds that remain undisturbed can persist in the soil seed bank for decades as they avoid some of the seed viability hazards previously described. Therefore, tillage slows the rate of turnover of the seed bank. In practical terms this could impact the rate of development of herbicide-resistant weed populations, with a slower shift in conventional tillage situations than under no-till. However, experimental evidence of this is lacking.

ii) **Tricking the weed seeds:** Many weed seeds, especially smaller-seeded annual broadleaf species, germinate in response to stimuli that indicate that the soil surface has been disturbed and cleared of competing vegetation. Light is the most common germination trigger, though many seeds also respond to temperature and moisture fluctuations, increased aeration, and increased release of nitrate and other soluble nutrients that occur in tilled soil. Many weed species also tend to germinate during specific times of year after a certain amount of soil warming. For these weeds, the time of emergence during spring can be approximately predicted by the accumulated growing degree–days (GDD), a summation of the number of degrees that daily average temperatures exceed a base temperature.

In the stale seedbed approach, the soil is tilled to prepare for seeding the crop, then planting is delayed for two or three weeks to allow a flush of weeds to emerge. Just before sowing the crop, emerged weeds are killed with no or minimal soil disturbance. Organic farmers can kill weeds by flaming or cultivating as shallowly as practical. Conventional farmers normally use herbicides at the end of the stale seedbed period, an option that may become open to organic producers in the future with the development of natural-product post emergence herbicides that are economically viable at the field scale.

In the false seedbed approach, weeds emerging in response to tillage are killed by two or more additional shallow cultivations at weekly intervals. The crop is planted immediately after the final cultivation. Because small weed seeds germinate better when the soil is firmed to enhance seed–soil contact, rolling is recommended after all except the final cultivation.

Adoption of SSB in severely infested areas, repeated tillage to exhaust soil seed bank and selective drying of panicles by direct contact of application of broad spectrum herbicides glufosinate ammonium, paraquat or glyphosate 15-20% concentration of the formulated product were found effective (Jose, 2015)

iii) **Conditioning weed seeds:** A different strategy is to avoid soil disturbance and other weed seed germination stimuli before and during spring planting. Keeping weed seeds dormant until the crop is well established can substantially reduce potential weed competition. This is a major objective in the no-till cover crop management and vegetable planting systems that some organic farmers and researchers are developing, in which winter annual cover crops are grown to maturity (heading in grasses, flowering in legumes and other broadleaf crops) then rolled or mowed to create an *in situ* mulch.

iv) **Soil solarization:** Soil solarization or "solar heating" is a non-chemical disinfestation practice that may serve as a component of a sustainable IPM programme. Solarization effectively controls a wide range of soil-borne pathogens, insects and weeds. Soil solarization is based on the exploitation the solar energy for heating wet soil mulched with transparent PE sheets to 40-55°C in the upper soil layer. Thermal killing is the major factor involved in the pest control process, but chemical and biological mechanisms are also involved.

v) **Herbicide application:** Herbicides are very effective at reducing weed populations and at the same time the number of seeds added to the soil seed bank. Weed seed bank densities tend to be greater in organic management systems than in systems reliant on herbicides, although this is not always the case as other factors such as crop rotation also strongly influence weed seed production. Pre-harvest applications of glyphosate can decrease seed production and impact seed viability in late flowering weeds. However, the slow action of glyphosate means that weeds must be managed well before the plant sheds its seed near maturity. The normal practice is herbicide application after one cutting, let the weeds to sprout again then go for herbicide application.

Advantages of Weed Seed Bank Estimation

1. **Predicting weed management efficacy:** Control Efficacy Knowledge of seed bank characteristics should be useful in predicting weed management efficacy. The composition of the seed bank regulates the timing and density of weed populations, both important components affecting control efficacy. In a field with a large initial seed bank,

metolachlor [2-chloro-N-(2-ethyl-6-methylphenyl)-N-(2-methoxy- 1-methylethyl) acetamide] controlled 75% of the giant foxtail in plots where heavy infestations (530 plants m^{-2}) had produced seed the previous year (Hartzler and Roth 1993).

Periodicity of germination and emergence of different weed species is also an important aspect of weed management. The time of weed seedling emergence influences which species will be the most serious weeds with a given crop production practice or most susceptible to certain control measures.

2. **Use of bio economic models:** A potential use of seed bank information in weed management is the use of bio economic models that incorporate seed bank dynamics and prediction of weed emergence into the decision making process A goal of these models is to incorporate weed population dynamics into weed management while accommodating multiple weed species and a wide range of control tactics. Model generated herbicide recommendations reduced weed control costs and resulted in an average annual herbicide application of 1.1 kg ha^{-1} compared to 3.5 kg ha^{-1} with a standard treatment (Forcella *et al.*, 1996)
3. **Economic Thresholds:** Economic thresholds for weed control have been criticized for not accounting for seed production by sub threshold populations of weeds on future weed management costs. An economic threshold is the weed density at which the cost of control equals the value of crop loss if no control action is taken (Cousens, 1987). Economic optimum thresholds (Cousens, 1987) incorporate the impact of seed production on future weed populations
4. **Management of Light:** Management of Light Exposure to light breaks seed dormancy in many plant species, including weed species. Because light can penetrate only a few millimeters in soil, dormancy may be induced in light-requiring seeds by shallow burial. A major source of light for buried weed seed is the light flash received during tillage. Increased weed seed germination after tillage in the light vs. darkness was first documented in 1969 (Wesson and Wareing 1969). Tilling the soil during darkness reduced weed populations in recent field research (Buhler and Kohler 1994), suggesting that eliminating or modifying light exposure of soil during field operations may have practical application for management of light-sensitive weed species. However, since all weed species do not possess light-sensitive germination, tillage in the absence of light will rapidly select for light-insensitive species.

Conclusion

Weeds continue to thrive in the face of increased efforts to eliminate them. Weed seed banks play a major role in this persistence. All forms of disturbance result in survival and selection of the best adapted plant species. Any cropping or weed control system that exerts a continuous, strong selection pressure will cause a buildup of the best adapted weed species and biotypes. Development of improved cropping systems will require an approach that concentrates on the pro-cesses and patterns linking scientific disciplines to agricultural systems. Agricultural systems are composed of inter-acting production, environmental, biological, economic, and social components. These interactions require the study of not only the parts, but also the whole system. Development of fully integrated and scientifically understood systems of weed management is a solution to ever-evolving weed problems and provides the basis for weed seed bank research.

References

Buhler, D. D. & K. A. Kohler. (1994). Tillage in the dark and emergence of annual weeds. Proceedings of North Central Weed Science Society, 49, 142.

Cardwell, V. (1987). Grain legumes as components in farming systems.

Cousens, R. (1987). Theory and reality of weed control thresholds. Plant Protection Quarterly, 2(1), 13-20.

Forcella, F., King, R. P., Swinton, S. M., Buhler, D. D., & Gunsolus, J. L. (1996). Multi-year validation of a decision aid for integrated weed management in row crops. Weed Science, 44(3), 650-661.

Hartzler, R. G., & Roth, G. W. (1993). Effect of prior year's weed control on herbicide effectiveness in corn (Zea mays). Weed Technology, 7(3), 611-614.

Jose, N. (2015). Biology and management of weedy rice (Oryza sativa f. spontanea) (Doctoral dissertation, Department of Agronomy, College of Horticulture, Vellanikkara).

Kropac, Z. (1966). Estimation of weed seeds in arable soil. Pedobiologia, 6, 105-128.

Luschei, E. C. (2003). Comparison of the effectiveness of seed bank sampling to seedling counts in reducing the uncertainty in estimates of weed population size. Aspects Applied Biology, 69,137-142.

Malone, C. R. (1967). A rapid method for enumeration of viable seeds in soil. Weeds, 15(4), 381-382.

Menalled, F. (2008). Weed seedbank dynamics & integrated management of agricultural weeds. Agriculture and Natural Resources (Weeds), 200-211. (Available online at: http://msuextension.org/publications/AgandNaturalResources/MT200808AG.pdf).

Menalled, F.D. (2013). Manage the weed seed bank-minimize "deposits" and maximize "withdrawals". (Available online at: https://articles.extension.org/pages/18527/manage-the-weed-seed-bankminimize-deposits-and-maximize-withdrawals).

Mesgaran, M. B., Mashhadi, H. R., Zand, E., & Alizadeh, H. M. (2007). Comparison of three methodologies for efficient seed extraction in studies of soil weed seedbanks. Weed Research, 47(6), 472-478.

Roy, D. K., Ranjan, S., & Sow, S. (2024). Efficiency and economics of broomrape (Orobanche spp.) control by different management practices in tobacco. Indian Journal of Weed Science, 56(2), 191-193.

Sahoo, S., Seleiman, M. F., Roy, D. K., Ranjan, S., Sow, S., Jat, R. K., et al. (2024). Conservation agriculture and weed management effects on weed community and crop productivity of a rice-maize rotation. Heliyon, 10(10).

Walia, U. S. (2014). Weed management. New Delhi: Kalyani publishers.

Walsh, M. J., & Powles, S. B. (2014). High seed retention at maturity of annual weeds infesting crop fields highlights the potential for harvest weed seed control. Weed Technology, 28(3), 486-493.

Wesson, G., & Wareing, P. F. (1969). The induction of light sensitivity in weed seeds by burial. Journal of Experimental Botany, 20(2), 414-425.

Wiles, L. J., Barlin, D. H., Schweizer, E. E., Duke, H. R., & Whitt, D. E. (1996). A new soil sampler and elutriator for collecting and extracting weed seeds from soil. Weed Technology, 10(1), 35-41.

Index